Lessons from Plants

Lessons from Plants

BERONDA L. MONTGOMERY

Harvard University Press
Cambridge, Massachusetts
London, England
2021

Printed in the United States of America

First printing

Credits for chapter opening illustrations:
Introduction, Chapters 1, 2, 3, 6, Conclusion: Morphart Creation / Shutterstock.com
Chapter 4: R. Wilairat / Shutterstock.com
Chapter 5: clipart.email

Library of Congress Cataloging-in-Publication Data

Names: Montgomery, Beronda L., author.
Title: Lessons from plants / Beronda L. Montgomery.
Description: Cambridge, Massachusetts : Harvard University Press, 2021. | Includes bibliographical references and index.
Identifiers: LCCN 2020047170 | ISBN 9780674241282 (cloth)
Subjects: LCSH: Human-plant relationships. | Plants. | Botany.
Classification: LCC QK46.5.H85 M65 2021 | DDC 581.6/3—dc23
LC record available at https://lccn.loc.gov/2020047170

To the memory of my beloved Dad
Roots of distinction bear great fruit

Contents

Preface

As a child I did not grow up on a farm or near a forest, the classic kind of upbringing that might explain my later enchantment with botany. But I did grow up in a house full of plants that were tended under the careful, observant eye of a mother with an incredible green thumb.

Plants were everywhere around us, both inside and outside our home. My mom's gardens were the gem of the neighborhood, a green oasis, a friendly abode for plants in a large city. She was able to cultivate such incredible flowers and healthy greenery because she was closely engaged with her beloved plants. She read and responded to their cues—this wilting plant needed more water, that one with the yellowing leaves required fertilizer, and the plant bending toward the light from the nearest window must be rotated so it could reorient itself. Mom's plant tending was part of her daily routine, an ingrained part of my childhood. She watched her charges closely in a way that can only be described as "listening" to them. When she noticed what they needed and supplied it, they responded by growing, and growing well. I can't say I fully understood the shared communication between my mom and her plants, but I noticed the beneficial outcome.

I did have a few of my own memorable encounters with plants, experiences likely similar to those of other children wandering outside in the sun-filled days of summer. Like most childhood memories, they revolve around danger . . . and eating. The care with which I watched out for poison ivy on long expeditions with my brother and sister. The plump, sweet wild blackberries I eagerly consumed on treasured walks during the lazy days of July. The nectar gleaned out of flowers stolen from my mom's carefully tended honeysuckle bushes, a sweet botanical discovery. I had no idea at the time that these delightful days experiencing plants in their environment would eventually lead to a fulfilling and significant career as a plant researcher.

In contrast, my natural aptitude for and love of science and math were obvious rather early in my life. While some of my family members found my obsession with quantitative and scientific topics odd, I often sought out "fun to me" activities like logic puzzles and homespun science experiments. I was never deterred by the fact that some of these experiments went awry . . . and may or may not have involved the local fire department! My interests were formally cultivated in middle school, when I had an opportunity to take advanced classes in math and science. Although my parents didn't understand where this burgeoning scientist came from, they were unwavering in their support—driving me to math classes at the local university after a full day of work, and faithfully taking me to the public library to do research and gather materials. Just as my mind and heart were beginning to function like those of a practicing scientist, I took a plant physiology course in college that set me on my path, turning my gaze fully to plant science. It was in that class that I got my first glimpse of the incredible science of plant life.

When I entered academic science as a biology researcher, I was prepared to experience many of the standard norms of researching science. I anticipated that I would form hypotheses and test them by means of probing questions and careful observation. I expected to conduct and oversee forward-thinking research, mentor future scientists, and, perhaps somewhere along the way, make some interesting, novel, and (hopefully) valuable contributions to what we know about how the world works. What I did not expect to happen, however, was the transformative growth and knowledge I gained from methodical and systematic observation of biological organisms, especially plants.

Following the plant physiology class, I embarked on my first official plant biology experiments. I explored a phenomenon by which the emerging leaves of some trees, including some species of oak, are transiently bright red in spring. After the first few weeks of growth, the anthocyanin pigments responsible for the red coloring are turned over and leaves appear their characteristic green color due to the accumulation of chlorophyll, the pigment responsible for driving photosynthesis. I conducted experiments in ecophysiology—the study of the interplay between the environment and a plant's physiology—to understand the purpose of this accumulation of red pigments. These investigations, which pointed to a role for the pigments as a sunscreen against ultraviolet light until the leaves mature, led to a decades-long, ongoing fascination with plants' responses to environmental light cues.

My enchantment with plants ultimately led me to pursue a career as a professor, with opportunities to continue doing research and teaching about these fascinating organisms. In both the classroom and the laboratory, I learned how important mentoring and leading were to my pursuit of success. Having not formally learned about either of these in my academic career, I began to seek out resources and opportunities to improve my skills as a mentor and leader in science, as well as the means to share insights with others in my community who are also eager to grow in these skills. My primary aim in this process has been to fully inhabit my space, life, and opportunities as I honor my goals and my own humanity, even as I ensure that I have the skills to support and value the humanity of those with whom I engage. The scholarly work that I have developed in studying and cultivating structures of support for effective mentoring and leading emerged from my careful observation of academic and scientific systems and their (mal)functions. As I have studied these systems, it has become clear to me that some of the biological principles that we recognize as contributing to the functioning of natural ecosystems hold great lessons for effective and equitable mentoring and leadership practices.

While many of us know a myriad of facts about plants' essential roles in supporting us, such as how they release human-sustaining oxygen and provide nourishment in the form of vegetables, nuts, and fruits, it is what plants do on their own, mostly not in relation to humans, that fascinates me most. Plants exist and thrive in so many seemingly uninhabitable places on the planet: the tree growing from a rock hanging over the ocean, shoots reemerging after a difficult Michigan winter, plants sprouting through what I thought was the impenetrable asphalt of my driveway. They have powerful, complex, and dynamic lives from which we can draw valuable lessons. As you will see, they survive and thrive in diverse environments, forge symbiotic relationships, collaborate, communicate, and contribute to their community.

I have learned a great deal about how to "be" in this world from my studies of plants. With this book I offer you a similar journey: to see how plants' individual and collective strategies and behaviors result in adaptable and productive living, and how we can learn from them. It is with this kind of knowledge and engagement that we, as humans, can better support ourselves and the other living beings around us.

Lessons from Plants

Humans have the least experience with how to live and thus the most to learn—we must look to our teachers among the other species for guidance. Their wisdom is apparent in the way that they live. They teach us by example.

—ROBIN WALL KIMMERER,
Braiding Sweetgrass

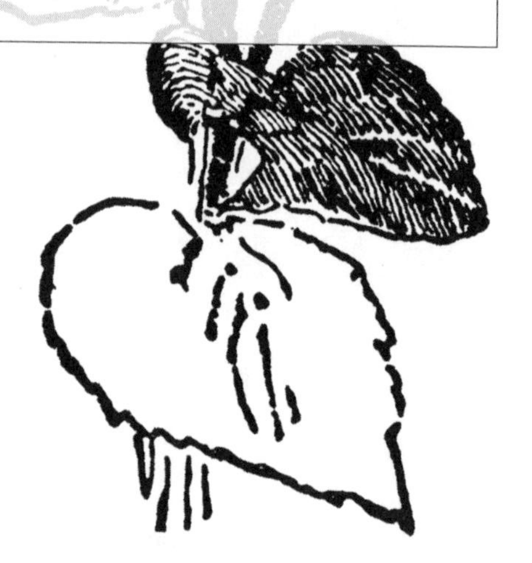

INTRODUCTION

A Sense of Self

Imagine a life in which one's entire existence must be tuned and tailored to the changing, and at times harsh, environment. A life in which there is no potential for escape. This is the life of a plant. It is difficult for us, as humans, to comprehend this kind of existence. Although we usually stay put in the face of short-term adversity because we have physiological mechanisms to deal with minor annoyances, like being too hot (sweating) or too cold (shivering), if such conditions persist or become more extreme, we can choose to uproot ourselves and physically move to a different, hopefully better, location.

Plants don't have that option.

Because plants are largely immobile throughout their life cycle, if they are to survive and thrive in dynamic environments, they must have a keen sense of what is going on around them and the ability to respond appropriately. From the very outset of life, sensing the environment is crucial. Where a seed lands and germinates determines the surroundings in which the resulting plant will spend its entire life.

Germination is the start of the life cycle for seed-bearing plants. The seedling emerges from the seed, and the plant then matures to adult stage. Following a period of vegetative growth, the plant enters the reproductive stage, when it produces flowers. The next stage progresses from flowering to developing seeds. After the mature seeds are released, the aging plant enters senescence, during which petals and leaves may be shed. In some species, individuals die after reproducing, while in others, they go through recurrent reproductive cycles.[1]

Although plants are all around us, most of us have little understanding of their exquisite abilities to anticipate, defend against, and adapt to constantly changing conditions. The failure to adequately recognize plants and their roles in the ecosystems we inhabit is sometimes referred to as "plant blindness."[2] This term has become increasingly controversial because it is based on a disability metaphor; that is, it reflects deficit-based thinking around blindness.[3] We might instead refer to the tendency to overlook plants as "plant bias." Indeed, experimental research and surveys

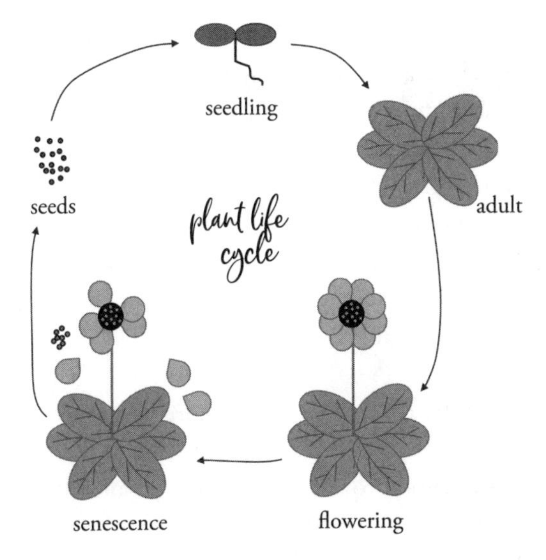

A seed-bearing plant starts life when a seed germinates, making the transition to become a seedling. The plant matures to adult stage and undergoes a second transition, to the flowering stage. The plant then progresses from flowering to developing seeds. After the mature seeds are released the aging plant enters senescence, during which petals and leaves may be shed.

have demonstrated that humans prefer animals to plants and are more likely to notice and remember them.[4] We also need a companion term that encourages a deepening awareness and appreciation of the plants around us: some refer to "flora appreciation," but I prefer "plant awareness."[5] Reducing plant bias and increasing plant awareness are important not only for plants, but for humans—for our physical, mental, and intellectual health.

The aim of this book is to increase your plant awareness, mitigate potential biases against plants, and introduce you to the wisdom of plants and what they can teach us.

One of the themes we will explore is how plants sense and respond to the environment. If you pay closer attention to plants around you, you will see many examples. You have probably noticed a houseplant stretching toward the light from a window. This plant is showing an active adaptation behavior—sensing and seeking out light. Because plants use light to produce their food (in the form of sugars) through the process of photosynthesis, they will bend themselves to get it.[6]

Another example is the dropping of a maple tree's leaves in autumn. This is a seasonal, energy-saving behavior; it would be expensive for the tree to maintain its leaves during the winter. Shedding its leaves allows the tree to persist in a quieter metabolic state. The brilliant, intensely hued burst of color that appears before the leaves fall (a result of the breakdown of the green pigment chlorophyll) is an illustration of the kind of complex behaviors that plants undertake in response to environmental cues.[7]

A maple tree's shedding of its leaves in the fall is different in one important way from a houseplant's bending toward the light. All plant species show some inherited adaptations, such as a distinctive leaf shape or a deciduous versus evergreen life cycle, that have evolved over time and are genetically fixed, being passed from one generation to the next. But plants also exhibit environmental adaptations that are not genetically fixed—rather, they occur within a single generation or lifetime and are usually not heritable. These environmentally determined changes are driven by which genes are expressed, or actively used. They include changes in a plant's phenotype (observable characteristics) such as leaf size, thickness, color, or orientation, or stem length or thickness, based on changing environmental cues. This kind of change of form and function in response to dynamic environmental conditions, such as light or nutrient availability, is known as phenotypic plasticity.[8]

Plants sense and respond to more than environmental conditions; their awareness extends to the plants and other organisms that surround them.

We could call them "nosy neighbors." Plants know "where" they are through environmental sensing, and they also know "who" is around them. That knowledge helps them make decisions about whether to collaborate or compete. They will compete with a neighboring plant for access to sunlight only if it makes sense to do so; if the neighbor is already significantly taller and the competition is unlikely to be successful, they will avoid competing. In some cases, as we will see, they may actually collaborate on gaining access to sunlight. Plants can also detect the behavioral responses of their neighbors, allowing them to extend their awareness of environmental cues and changes. And they sometimes even change their behavior depending on whether their neighbors are relatives.

Plants receive and respond to both internal and external cues and appear to have a recognition of ecosystem diversity, that is, they can perceive the range of individuals around them and the responses these neighbors exhibit to environmental cues. They monitor external changes and initiate internal communications pathways to coordinate their response to dynamic conditions.[9] The cues they respond to can be abiotic, or nonliving, cues, such as information about the temperature or the availability of light, water, or nutrients. Biotic signals, those originating from other living organisms, also serve as potent cues that can, for example, enable a plant to mount a defense against predation, herbivory, or bacterial or viral infection. Some plants, when attacked by insects, produce compounds that inhibit the digestion of the attacking insect, thereby limiting further damage.[10]

Plants may even have a form of memory. In some cases, it is mediated by epigenetic changes. Epigenetic changes modify how genes are expressed or activated; they do not alter the genetic code itself. An environmental stimulus may cause a molecular "flag" of sorts to regulate whether or not a gene is used to produce a protein. The change in protein regulation then modifies the plant's phenotype. Such epigenetic changes are sometimes transmitted to subsequent generations. The definitive mechanisms and specific roles of the environment in transgenerational epigenetic control in plants are still under study.[11]

One of the best known examples of plant memory is vernalization: certain plants will not flower until they have been exposed to a lengthy cold period. The winter cold is "remembered" as a sign that the plants should flower in the spring. Sun-tracking plants, such as sunflowers and Cornish mallow, also display memory, turning toward the direction of the sunrise before dawn.[12]

Plants use internal and external cues alongside adaptive behaviors and energy budgeting to make the most of the environment in which they grow. Photosynthesis requires light, inorganic carbon (in the form of carbon dioxide), and water, and plants also need nutrients like phosphorus and nitrogen. Therefore, it is not surprising that they are extremely sensitive to the availability of these resources and manage their energy budgets carefully. To make their food, plants allot energy to grow the leaves needed for harvesting sunlight. Then they convert the gathered light energy to chemical energy (sugars), using carbon dioxide and water. At the same time, they limit nonproductive uses of energy. In favorable light conditions, for example, they contribute energy to leaf building while diverting energy away from stem elongation.

Plants also show finely tuned adaptive responses when nutrients are limiting. Gardeners may recognize yellow leaves as a sign of nutrient deficiency and the need for fertilizer. But if a plant does not have a caretaker to provide supplemental minerals, it can proliferate or elongate its roots and develop root hairs to allow foraging in more distant soil patches. Plants can also use their memory to respond to histories of temporal or spatial variation in nutrient or resource availability.[13] Research in this area has shown that plants are constantly aware of their position in the environment, in terms of both space and time. Plants that have experienced variable nutrient availability in the past tend to exhibit risk-taking behaviors, such as allocating energy to root elongation instead of leaf production. In contrast, plants with a history of nutrient abundance are risk averse and conserve energy. At all developmental stages, plants respond to environmental fluctuations or unevenness so as to be able to use their energy for growth, survival, and reproduction, while limiting damage and nonproductive uses of their valuable energy.[14]

Altogether, these types of responses suggest that plants are able to learn and remember, if we understand learning as a change in behavior based on active recall, and memory as cellular communication about prior experiences.[15]

Because they exhibit a kind of awareness and memory, one could consider that plants know "who" and "what" they are. They proceed from this knowledge of self to go about *being.* And it is in the process of being that plants discern, respond to, *and* influence patterns in the environment. In other words, plants give survival their very best effort, while fully assessing the potential for success based upon the specific environment in which they exist.

So, while it may indeed look to the uninformed eye as though plants are just "sitting there," they exhibit awareness and intelligent behaviors from the very earliest stages of development until senescence or death. They have developed extraordinary abilities to sense what is going on around them and tune their growth and development to environmental cues to maximize productivity and survival. Because of this constant exploring and monitoring, plants should not be viewed as immobile and passive, argues the philosopher Michael Marder; the place occupied by a plant "dynamically emerges from the plant's living interpretation of and interaction with its environment."[16]

Whether one understands plants as aware or as intelligent, behind either concept lies a general appreciation of plant behavior. The idea that plants "behave," rather than passively existing or growing, has only recently become more widely accepted among biologists. Behavior in plants often manifests itself in the way they grow—growing at a different rate, or in a certain direction. Because plants grow slowly, their activity occurs on a different time scale from the kind of movement we call "behavior" in animals.

Another obstacle to accepting the idea of plant behavior came from the long-standing belief that behavior is only possible in organisms with a central nervous system, which plants lack. But scientists began to understand behavior more broadly, as describing the ability to gather and integrate information about the condition of the external and internal environment and then using that information to alter internal signaling or communications pathways (neural networks in animals and signal transduction pathways in organisms such as plants), resulting in changes to growth or allocation of nutrients and other resources. With this understanding, the idea that plants can "behave" became more acceptable.

Once we acknowledge that plants exhibit behavior, does that mean they are also able to "choose," "make decisions," and have "intention"? Most plant scientists agree that the ability to distinguish between multiple signals and to selectively alter behavior based on one signal over another is evidence of decision-making. Plants also have intention, contends Michael Marder, though it differs from intention in animals: "When animals intend something, they enact their directedness-toward by moving their muscles; when plants intend something, their intentionality is expressed in modular growth and phenotypic plasticity. Plant and animal behaviors are the accomplishments of the goals set in their respective intentional comportments."[17]

The next question, whether these abilities demonstrate that plants possess intelligence or consciousness, is a topic that garners both avid supporters and a larger group, perhaps, of detractors. And others still remain agnostic, noting that plants need not have consciousness or intelligence to be viewed as worthy of study and awe.[18] Regardless of whether plants possess awareness—the ability to perceive what is going on around them and respond accordingly—and consciousness—the ability to actively perceive, contemplate, and assign meaning to a decision about a particular response—the complexity of plants and their abilities to sense, integrate, and respond to environmental stimuli has been increasingly accepted. Further, whereas there are still controversies about, and in some instances resistance to, seeing plants as intelligent, there is increasing consensus that plants, and other organisms such as ants and bees that lack highly developed brains, can exhibit intelligent behaviors that allow them to respond as individuals or in community to a dynamic environment.

The evidence that plants make adaptable choices—behaviors that increase their success and persistence—

merits deep reflection and can provide valuable lessons for humans. Like all biological organisms, plants usually make choices that are clearly beneficial, yet they can also initiate behaviors that we might characterize as bad, either because maladaptive for the plant itself, or because harmful to others. Biologists believe that, with some exceptions, the choices a plant makes will usually benefit its own survival and reproduction because over evolutionary time, plants that make better choices will have more offspring than those that make worse choices. But sometimes what's good for one species is bad for another. For example, some plants can harm neighbors through release of chemical compounds or by taking over entire ecosystems. The latter strategy often characterizes plants deemed invasive, such as kudzu, which is a major ecosystem problem in the southeastern United States, as it has replaced native plants and impacted local insects and other animals.[19]

Despite the harm that plants can sometimes do, for the most part, their behaviors work to benefit their own flourishing and that of the communities in which they live. In the pages to come, we will explore many of those behaviors. We can learn a great deal from observing how plants live in their environment. Notably the knowledge of plants—lessons from these organisms on *being*—shows us that you thrive or languish based on your ability to know who you are, where you are, and what you are supposed to be doing. Then you must find a way to carry on from this "sense of self" to your surroundings and to pursuing your purpose, a task that may be challenging if you are in distress, compromised, or have mutated from your ingrained, encoded, or adapted purpose. Plants in distress have some means to improve their chance of recovering from stress and resuming growth. And if the plant has a caretaker with the ability to recognize signs of distress, that caretaker can provide the necessary assistance.

All of the activity that plants engage in—operating sophisticated light catching systems, foraging for nutrients, communicating warnings of danger to others within their community—is how plants sense and adapt to their environments. It is how they survive and thrive. And it is occurring all of the time, right in front of us.

As humans we must first pay attention. We must look beyond what is quickly observed to be fully aware of how plants support themselves and the other organisms with which they live, and how they transform their environment. Then, after careful, close observation, we must ask the right questions to learn from them about how to live with purpose, agency, and intention. And maybe we can take on some of these behaviors. Their lessons are ours for the learning.

There is no question that plants have all kinds of sensitivities. They do a lot of responding to an environment. They can do almost anything you can think of.

—BARBARA McCLINTOCK,
quoted in EVELYN FOX KELLER,
A Feeling for the Organism

1

A Changing Environment

I vividly remember one of the first science experiments I ever performed, when I was in kindergarten. By watching a simple bean seedling grow, I learned about the remarkable ability of plants to adapt to their environment—and now, decades later, I am still in awe at that ability. The experiment was coordinated by my kindergarten teacher, who instructed each of us to grow a bean seedling on a windowsill at home. We were to put wet cotton balls or some wet soil in the bottom of a plastic cup, add a few beans, and observe them daily. One day when I looked at my beans, I made an exciting discovery. I noticed that a crack had appeared in one of them, and a tiny root was emerging from the crack. Then, in the days that followed, a stem began to emerge from the other end of the bean, and tiny leaves unfurled. Reaching toward the sun in our window, the bean plant continued to grow.

A few weeks later, the teacher asked all of us to bring in our seedlings for show and tell. The plants, I was surprised to see, were not all the same: some were short and stocky, while others were tall and spindly. The teacher explained that these differences depended on how much light we'd each had coming through our window. If the windowsill was shady, the plant would grow tall to try to reach the light. This was my first exposure to an essential feature of plants—that they are exquisitely attuned not just to light levels, but to a whole array of environmental conditions.

Plants are aware of light, water availability and moisture level, and nutrient abundance in the soil. They perceive changes in these factors as they scan the environment and assess what responses they need to make. Based on the information they gather, they are able to alter their behavior, morphology, and physiology in response to changes in their surroundings.

Most of us know that bean seedlings, like other green plants, use light to make food through the process of photosynthesis. But few of us know the fascinating details of how they respond to shifting light conditions. Light influences plants from the very beginning of their life cycle; while still underground, some seeds are stimulated by light to germinate.[1] While the root follows gravity to grow downward, the shoot grows upward toward the light. The first leaves to appear are the embryonic leaves, or cotyledons. They accumulate molecules of the pigment chlorophyll, which "captures" light energy. The bean seedling's leaves appear green to the human eye because chlorophyll absorbs red and blue light, leaving the green portion of the visible spectrum to pass through or be reflected. The photoreceptors in our eyes see the wavelengths that are not used by light-gathering photosynthetic pigments.

As the seedling continues to grow and mature, its leaves stretch toward the sun to gather photons—quanta of electromagnetic energy. Chlorophyll molecules in the leaves convert light energy to chemical energy. That energy is then used to turn carbon dioxide into carbohydrates. It is through this process of photosynthesis—the harvesting of sunlight to drive conversion of inorganic carbon, in the form of carbon dioxide, to fixed carbon, in the form of sugars—that plants make their food.

The bean's new leaves are not just passive recipients of light. They make adjustments depending on how much light they receive. But how do they measure the light? Scientists have discovered that plants are able to detect the number of photons absorbed by a unit of leaf surface area per unit of time. The rate of photons hitting the surface of a leaf affects many plant processes because it controls the rate of photosynthetic reactions; more photons means more excited electrons, which means faster reactions.

The chlorophyll molecules that are central to this calculation of photon density are contained in complex light-gathering systems, called "antennae," that trap and transport light energy to "reaction centers," where the chemical reactions take place. The efficiency with which plants collect, convert, and harness energy can easily rival any solar cell. But the bean plant in your garden can do something no solar cell can currently do—it can modify its light-gathering structures in response to dynamic external cues such as dim versus bright light, or a change in the predominance of different colors of light.[2]

Experiments that my laboratory and others have conducted with plants and cyanobacteria—bacteria that carry out photosynthesis—reveal a remarkable ability to adjust the light-gathering system to adapt to different light conditions. If light is too dim, levels of photosynthesis can be too low to provide the organism's energy needs. But too much light exposure is also detrimental. When available light exceeds the capacity for light absorption, the excess energy can generate toxic byproducts. What a plant wants to do is to maximize light absorption while limiting damage. It does this by "tuning" its light-harvesting system to external light conditions.

Plants and photosynthetic bacteria tune their antennae in several ways. They are able to match the specific light-harvesting proteins contained in the antennae to the wavelengths of light available. They can also adjust the size of their light-harvesting complexes; these complexes become larger in low light conditions, to increase light absorption, and smaller in bright conditions, to limit potential damage. It is an intricate balance to obtain just enough but not too much light energy. Through these complex modifications of their light-gathering system, plants maximize their energy production to support essential activities.

At the same time that a newly sprouted seedling is making these adjustments within the cells, it is also adjusting its stem and leaves in an effort to maximize light absorption. The difference in height of the bean seedlings that my fellow kindergartners and I had brought to class was the result of coordinated communication between the seedlings' tissues and organs based on available light. Stem position is vitally important, since it determines the location of the leaves, and it is the leaves that absorb the light required for producing chemical energy and sugars. When the leaves sense that they are in a favorable position for receiving adequate light, they send a chemical "stop" signal to the stem, which inhibits further elongation. This process, known as de-etiolation, results in plants with short stems and well-developed leaves. If the leaves are not able to harvest enough energy because of poor light conditions, however, they send a "go" signal to the stem to elongate, with the goal of getting the leaves into better light. This process, etiolation, results in seedlings with long stems and few leaves.[3]

Such a coordinated response between stems and leaves is a powerful example of how plant organs communicate in response to changing environmental cues. Plant scientists are becoming increasingly aware that sensors that detect these cues, including light-sensitive receptors, regulate this kind of interaction.[4] For example, my research team's investigations yielded insights into the roles of specific genetic signals used for communication between leaves and stems to regulate de-etiolation, as well as the roles of signals from both shoots and roots in light-dependent regulation of root development.[5] Scientists use the term "developmental integration" for the idea that the integrated functions of an organism depend on coordinating the activity, development, and functions of each individual part.[6] This kind of integrated response is essential for our bean seedling. It cannot uproot itself and move to a better location to escape drought or shade;

instead, it responds to a whole host of stop and go signals that trigger physiological and structural changes to improve its situation. Such developmental plasticity is critical if a plant is to survive in a dynamic environment.

In the most extreme case, a bean seedling can even survive for a time with no light at all. Scientists who have observed plants growing in the dark have found that they are vastly different in appearance, form, and function from those grown in the light. This is true even when the plants in the different light regimes are genetically identical and are grown under identical conditions of temperature, water, and nutrient level. Seedlings grown in the dark limit the amount of energy going to organs that do not function at full capacity in the dark, like cotyledons and roots, and instead initiate elongation of the seedling stem to propel the plant out of darkness.[7] In full light, seedlings reduce the amount of energy they allocate to stem elongation. The energy is directed to expanding their leaves and developing extensive root systems. This is a good example of phenotypic plasticity. The seedling adapts to distinct environmental conditions by modifying its form and the underlying metabolic and biochemical processes.[8]

Plants exhibit phenotypic plasticity in response to many environmental conditions, not just light availability. They can respond to stresses such as drought, variations in temperature, or lack of space and nutrients.[9] To maintain a constant yield of seeds under different conditions, for example, the bean plant can modify any of several components of plant yield: the number of pods, the number of seeds per pod, or the size of individual seeds.[10]

The kind of phenotypic plasticity that results in irreversible adaptations is known as developmental plasticity. These changes, which occur during the plant's development or affect vital processes, are often visible. We may observe that roots or stems become elongated, leaves stop being produced, flowering occurs at a different time than usual, or seeds are smaller.

In contrast, physiological, or biochemical, plasticity refers to reversible adaptations that happen inside the cells.[11] Because it does not produce easy-to-observe changes like a stem bending toward the sun or leaves changing color, this type of plasticity is easy to overlook. But it is equally important, enabling a plant to tune its light-harvesting complexes to respond to different light levels, or alter the ratio of different photosynthetic enzymes in response to carbon dioxide levels to ensure that light energy is being used productively.[12]

Underlying the need to adjust its form and metabolic processes to the environment is the bean plant's energy budget. The seedling has a certain amount of energy that must be used for maintaining daily activities, but it can be divided up in different ways. Should more energy go toward building a new leaf, or lengthening the stem? Elongating roots, or forming flower buds? These questions are much like those that go into our monthly financial budgets. After paying for rent, I see how much money I have left for food. Will it be ramen noodles, or sushi with friends for lunch? If I need to make a major new purchase such as a car, it may be ramen noodles—for several months. Ultimately, if I do not have enough money for essentials, I need to work longer hours—in the same way that the plant needs to adjust to absorb more light energy. A plant's ability to adjust its energy budget to a changing environment is critical for its survival.

All living things have energy budgets, but they manage them in different ways.[13] Animals adapt by changing their behavior and regulating their movements. In temperate climates, for example, bears and other animals hibernate over the winter to save energy when food is scarce.[14] Plants adapt in a different way. As we saw with the bean seedling, they can change their form, or they can make biochemical changes. Some plant biologists consider both of these modes as types of behavior.[15] Another difference between plants and animals is that plants modify their behavior and form in response to the environment at different developmental stages.[16] While a seedling may elongate its stem or grow more leaves in response to light availability, a mature plant may find that it needs to reposition its leaves. Leaf position can be modified by changing the water pressure, or turgor, within the cells, or by causing different parts of the petiole—the stem of the leaf—to grow at different rates. On a scorching summer day, for instance, a plant can lift its leaves off the surface of dangerously hot soil.[17]

Different considerations come into play for a tall oak tree. Up in the canopy, some leaves might not receive sufficient light because they are shaded by other leaves, or they may receive different wavelengths of light.[18] By bending or lengthening their petioles, those leaves can move into open spaces with more or higher quality light.[19]

When we think of environmental conditions, we usually think of light, water, nutrients, and so on. But the bean plants and lilac bushes in your garden contend with another environmental factor, too: the rabbits and deer that feast on them. Gardeners and horticulturalists are well aware of what biologists call animal-induced plasticity, which occurs when an animal nips off a branch or stem and new lateral, or side, branches emerge. Sometimes we initiate this response ourselves, by pruning. We prefer bushes with the compact appearance that arises when new lateral branches are initiated, rather than the spindly and sparsely branched bushes that grow naturally in the wild.[20] But the bushes may have evolved this response for a reason: a more dense form may make it more difficult for animals to gain access to flowers and fruits.

In addition to developmental and biochemical plasticity, plants can also respond to the environment by means of epigenetics. Recall, from the Introduction, that epigenetic changes are changes that affect how DNA is regulated, and some of them may be heritable. One process that is subject to epigenetic regulation is vernalization. Vernalization refers to the promotion of flowering after exposure to a long period of cold—the result being that flowers do not bloom until chilly winter gives way to spring. Scientists have discovered that cold temperatures modify gene expression in plants with this trait, and that the modification is maintained for months, through myriad cell divisions. The plant is effectively able to "remember" that it has passed through winter and that it is safe to flower. This memory does not persist to the next generation, however.[21] There is some evidence that certain plants, such as the valley oak (*Quercus lobata*), may experience epigenetic changes in response to climate change that are passed down to the next generation.[22]

So far we have been discussing leaves, stems, and branches—all structures that are above the ground. But plants respond to environmental conditions below the ground as well, where there is competition for limited resources.[23]

Soil conditions are far from uniform. The pH level can vary from place to place; decomposing leaves or an animal carcass may create a nutrient-rich patch.[24] This patchiness can also result from the uptake and resulting depletion of resources by other plants or by soil microbes.[25] Hidden from view, plant roots are able to detect unevenness in the availability of water, minerals, and nutrients in the soil.

In the case of poor soil conditions, plants can contribute more of their energy budget to root development. The roots branch and develop root hairs—long, thin roots—that reach down and out to search for nutrient-rich, well-watered soils.[26] In nutrient-rich patches of soil, plants are able to increase root biomass. Roots grow to the side and increase in density to take advantage of the good conditions. Roots respond to temporal as well as spatial variation. Plants can increase their root biomass if more nutrients become available in the short term.[27]

These changes in root structure and growth are usually initiated and promoted by hormones. The primary hormone used is auxin, the same one that is involved in causing plants to bend toward the light.[28] Root modifications also have wide-reaching consequences, affecting the above-ground parts of the plant, too. When nutrients are limited, plants shift energy away from shoots and toward roots, as well as toward transport proteins that are involved in nutrient uptake.[29] But when nutrients are abundant and roots are taking in healthy amounts of essential nitrate, which is used for the production of proteins and other critical cellular compounds, the hormonal balance shifts and promotes more branching of shoots.[30]

Next time you are in your garden or walking in the woods, spare a thought for all that is going on underground. The ability of the bean plant and the oak tree to control root initiation, growth, and density is crucial for supporting the plant's growth and reproduction.[31]

Plants, as we have seen, have an extraordinary ability to sense conditions in their environment and respond to them. Humans can learn some useful lessons from them that can help us thrive as individuals and a community. Just as the bean seedling detects exactly how much light is striking it and which nutrients its roots are absorbing, we need to be acutely aware of our surroundings, reflecting intentionally on what we perceive and how best to respond. Do we have adequate food and shelter? What about emotional, financial, and logistical support from our family, friends, and workplace? These are questions that we must ask in both the short and the long term. Whereas we may have long-term plans in place to support our basic needs, we can be subjected to sudden changes or disruptions to our plans that require us to respond in the moment.

One of the greatest lessons I have learned in this regard is the importance of intentional self-reflection, or the equivalent of taking time to perceive my environmental conditions. It's not uncommon for me, and other humans, to be in a constant state of busyness with little time reserved for self-reflection, little time taken to assess whether our current actions still align meaningfully with our current environment. The importance of prioritized reflection time to sense conditions, stay in tune with my environment and the available resources and support, and then proceed in responding accordingly is what I've come to understand as a need to "process and proceed."[32] Such functioning is similar to plants' environmental responsiveness.

Based on its circumstances at a particular time, the bean plant may decide to grow taller or extend its roots, putting more of its energy budget toward one structure or another. In the same way, we need to make a strategic plan for how much energy to allocate to which activity and determine where within our community to search for greater resources depending on current conditions. We might realize that to support ourselves and our basic needs we need additional resources, in which case we may need to ask for a raise, move, or take a class instead of eating out.

Sunlight and nutrients are not static, and neither are the circumstances of our lives. When a situation changes, it's important to be aware and to respond accordingly. The humble bean seedling provides an excellent example of how to adjust and readjust to outside circumstances.

The trees and plants show respect for each other by the way they live in harmony.

—MASARU EMOTO,
The Hidden Messages in Water

2

Friend or Foe

Gardeners sow seeds with visions of a yard full of colorful blooms, or a bountiful harvest. We welcome the emergence of seedlings in the spring. But we don't just toss a random set of seeds into the soil. Experienced gardeners think carefully about which flowers and vegetables to plant and how to group them. The most astute of gardeners pick "friendly" species that grow well together to ensure a healthy, collaborative environment, and we often give the seedlings a head start by germinating them inside. We carefully adjust the light, humidity, and watering schedule to nurture our young zinnias, beans, and tomatoes. After the danger of frost is over and we transplant the seedlings outside, there is still more work to be done. In the weeks that follow, to maintain a healthy garden, we begin culling. Thoughtfully and carefully, after assessing the spatial distribution of the young plants, we sacrifice a few so that the rest have plenty of room to grow and won't compete with each other for sunlight and nutrients. We weed, too, removing unwanted species like dandelions and ragweed.

In a natural environment, culling takes place without the hand of the gardener—some seedlings fail, some are consumed by herbivores, and others grow to adulthood. You might observe a tight cluster of oak seedlings and imagine a brutal struggle for survival. But there is much more going on. The competition between the seedlings is tempered by their judgment about how much energy to expend—and there is collaboration going on as well as competition, in ways that might surprise you. The zinnias, beans, tomatoes, and oaks are constantly assessing whether neighboring plants, insects, fungi, and bacteria are friends or foes, and making choices about how best to focus their energies to gain needed resources.

As we have seen, plants are budget conscious regarding energy investment. For this reason, they avoid competing more than necessary; they will compete with neighbors, say, by trying to gain access to light or nutrients, only until they have met their needs. They use a number of different mechanisms to gauge when to initiate competition, and when to disengage—to put on the brakes—to avoid using valuable energy resources unnecessarily.[1] If they persisted in competing for resources after they had obtained enough, they would be drawing down energy that they might need in the future. Rather than competing, plants may choose to collaborate instead; this allows them to save energy by sharing acquisition costs.

Deciding how to interact requires a plant, such as an oak sapling, to be aware of what other organisms, whether plants, animals, or microorganisms, are present, whether it can communicate with these neighbors to assess common or complementary needs, and whether any mechanisms are available to work synergistically on acquiring resources while sharing the cost.

How does the young oak assess whether a neighbor is a friend or foe, and what to do about it? Botanists believe that many species use what is known as the detection-judgment-decision paradigm, a model that was first developed by animal behaviorists. Something as simple as a honeybee visiting a flower may seem like a random act to us, but in fact it begins with specific detection of a flower by the bee. Flower detection is followed by a judgment on the part of the bee about distinguishing between flowers, and ultimately a decision to actively engage.

Indeed, scientists have found that honeybees exhibit cognition that allows them to distinguish between visual cues, even very similar ones. Based on rewards such as sweet nectar, compared to no reward or a penalty such as a bitter substance, bees can use visual cues to discriminate and make decisions about flower selection.[2]

The process works the same way in plants. The plant uses its receptors to detect information, which may lead to production of a cue such as an electrical signal or transient calcium or hormone accumulation; it processes and evaluates this information, often by means of detected hormones, to make a judgment; and it then makes a decision, such as whether to change its phenotype by altering gene expression.[3] As an example, consider studies scientists have done on thale cress (*Arabidopsis thaliana*), a small plant whose leaves grow in a rosette, like a dandelion. When many cress plants are growing close together, their leaves may be shaded by those of their neighbors. The researchers found that the plants are able to detect the close proximity of neighbors when their leaf tips touch. This signal is perceptible even before the plant notices changes in the light spectrum associated with crowded leaves.[4] Using this information about its neighbors, the plant can make a decision about how to gain access to more light.

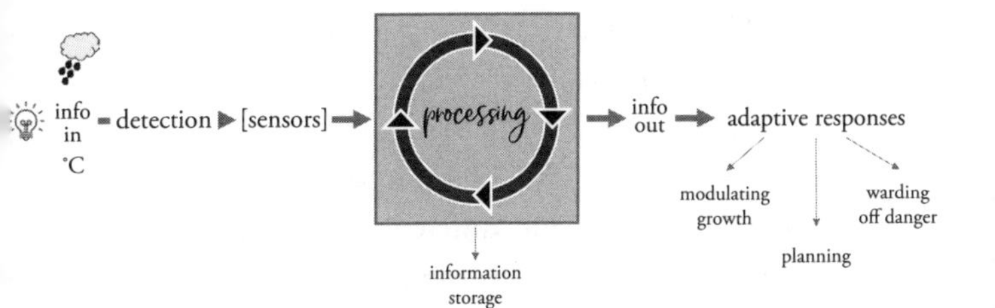

Plants employ a decision-making process that is based on their ability to detect environmental cues using sensors (e.g., light, temperature, and moisture sensors), to process the information they take in, to store some of it (e.g., in the form of epigenetic changes), and to initiate adaptive responses, which include planning, modulating growth, and warding off danger.

Many plants also detect the presence of each other through volatile organic compounds (VOCs) that are released into the air. These compounds are secondary metabolites that are not directly used for growth, development, or reproduction, although they can induce or interact with hormones that do control these processes. They are often regarded as a form of language.[5] Biologists used to think that only animals had the capacity for self-recognition and kin recognition—that is, the ability to tell if a tissue or another individual is genetically identical or closely related to oneself. But experiments have shown that plants, too, can recognize self and non-self, as well as kin.[6] This recognition is most often mediated by the VOCs that plants produce either routinely or in response to a specific environmental signal, like a beetle chewing on a leaf. These compounds serve as a means of communication from one plant to others, both of the same and different species, and even from a plant to individuals in other groups, such as insects and bacteria.[7]

In deciding whether to compete or collaborate, a plant carefully weighs the opportunity costs of investing its resources into one function rather than another.[8] Just like we do when determining the best option for ourselves—should I look for a job now, or pursue additional education first?—plants decide on a course of action when the expected benefit of that action outweighs the cost. As you'll see, this ability to respond dynamically to the environment has both short-term and long-term benefits.

The young oak tree, like all plants, needs light, nutrients, and moisture to grow, develop, and reproduce. These resources can be in short supply, though, especially in crowded conditions. When resources are limited, those individuals that are better able to access them, or can use them more efficiently, will survive better and have more offspring. It is possible that plants evolved these responses due to natural spatial variability in the availability of resources in the environment, or use by other organisms in a niche, for example, nutrient utilization by soil-dwelling organisms such as microbes.[9] The oak tree that has these abilities, as well as mechanisms to avoid damage from predators, will persist to pass more genes on to the next generation, and the strategy will spread.

When in a head-to-head situation, plants generally choose one of several responses: confront and compete, collaborate, tolerate, or avoid altogether. In response to environmental cues, they respond in the most budget-conscious way.[10]

A classic example of confronting and competing involves the rivalry for light. Because of their requirement for sunlight, plants are extremely sensitive to the presence of neighbors, which may shade them from the rich red wavelengths that are present in full sunlight and drive maximal photosynthesis. Shady neighbors limit photosynthetic potential and the production of chemical energy.

Plants respond to this scenario by engaging in a form of competition that is known as shade avoidance behavior.[11] When the leaves of an oak sapling are being shaded by its neighbors, the production or accumulation of growth-promoting hormones is activated.[12] Those hormones cause the stem to elongate (a form of developmental plasticity), and the shaded sapling can "win" by growing taller than its neighbors. The sapling may engage in competition for sunlight with nearby neighbors by "racing" to an opening in the canopy or another area with direct exposure to the sun. Other strategies include tilting the leaves up, reducing branching while sending more resources to the main stem, and increasing root growth.[13] The winner of this race is able to replenish and increase its energy stores, which it can use to increase the production of biomass, defend itself against danger, or reproduce.[14]

In Chapter 1, I mentioned the stop and go signals that tell a bean seedling when it should stop or start elongating, depending on how much light it is receiving. A similar process occurs when an oak sapling interprets environmental signals to determine the threat of competitors and initiate a response.[15] An array of proteins called photoreceptors are able to detect light of different wavelengths.[16] They tell the sapling not only how much light it is receiving, but what the quality of that light is. When photoreceptors detect a large proportion of light in the far-red region of the spectrum they send a go signal, prompting the plant to change position to reach more direct light. Far-red light, just at the border of the wavelengths visible to humans, indicates the kind of poor quality light that is typical of shade. But if the receptors detect a large proportion of red light, typical of full sunlight, they signal plants to stop elongating the stem because they are in a good environment to inhabit. Such checks and balances in the communication system allow the sapling to respond expediently by changing its phenotype while reserving resources for other activities that will increase growth, persistence, and reproduction.

Another way plants compete for light is through lateral, rather than vertical, growth. In this strategy, instead of growing upward, plants grow to the side into gaps that are more open.[17] Lateral leaf growth is a much more complex process than was first thought. Researchers have made the remarkable finding that some plants may adjust their competitive or collaborative behavior depending on whether their neighbors are close kin or not. Such behavior is well known in animals and is thought to have evolved because kin share genes. A blue jay, for example, shares on average half its genes with its siblings, and some of these genes are linked to increased survival—so if the bird protects its sister or brother from a predator, it is also ensuring that its own survival-enhancing genes persist.[18]

Now, researchers have discovered something similar going on in plants. Plants, they have found, may use lateral leaf growth for collaboration rather than competition. Studies on yellow jewelweed (*Impatiens pallida*) have shown that individuals are able to recognize closely related kin through their roots. Those that are planted next to siblings grow differently from those that are next to strangers. Rather than competing for light, they collaborate. Plants growing next to kin branch more and become bushier—as a result, they reduce leaf overlap and shading of their related neighbors.[19] Other experiments with thale cress have also shown a reduced competitive response in plants whose neighbors are kin. In this case, the plants may recognize their kin by aboveground rather than belowground signals, mediated by light-detecting photoreceptors.[20]

Such behavior has been observed in a variety of other species, including trees. Next time you are in a forest, look straight up toward the sky, or if in an airplane look down on a tree grove from above. You may notice gaps between the crowns of neighboring trees. This phenomenon, called "crown shyness," or "crown spacing," was initially thought to be driven largely by abrasion due to physical proximity.[21] More recent work, however, has shown that it may be the result of photoreceptor-mediated shade avoidance or, in some cases, collaborative behavior. Crown spacing is more common among closely related trees than among nonrelated trees of distinct species.[22] So it appears that plants compete for access to light less frequently with kin, or closely related plants, than with non-kin. Careful spacing in the canopy is a collaborative developmental response that limits competition and is an example of how plasticity responses to competition can impact ecosystem and community level dynamics and can determine, ultimately, which species persist.[23]

In addition to competition and cooperation, plants in light-limited conditions sometimes respond with tolerance. In this case, plants don't race for the light through hormone-mediated growth and reallocation of resources as they do when practicing shade avoidance.[24] Rather, shade-tolerant plants initiate adaptations that allow them to produce sufficient food in conditions with limited light. They have thinner, larger leaves with greater concentrations of chlorophyll pigments to capture more of the red light that is in short supply in dim conditions.[25] As a trade-off, shade-tolerant plants spend less energy making the pigments that serve as sunscreen in sunny conditions. Thus, shade-avoiding plants and shade-tolerant plants display adaptations that allow them to optimize light capture and fitness: shade-avoiding plants optimize for sunny conditions, while shade-tolerant plants optimize for shady conditions.

As an oak sapling or jewelweed seedling grows, it senses its neighbors and assesses their proximity, size, and relatedness through various means, both aboveground and belowground. Based on what it finds in its immediate environment, it then makes the decision—in terms of molecules synthesized and deployed—whether to engage in competition, cooperation, avoidance, or tolerance.[26] The ability to determine when to compete and when not to is critical to a plant's energy decision-making, allowing it to use its resources in the most effective way.[27]

Plants positioning their leaves and stretching tall to gain access to light may be visible to the observant human eye; but there is an active battlefield underground, too. Just like leaves, plant roots compete for physical space and resources.

You may think of roots as uninteresting, but they come in as wondrous a variety of sizes, lengths, and arrangements as do blossoms, leaves, and stems. Some plants' roots are shallow and highly networked, with many branches and cross-connections, while others are long and deep, with taproots that probe the subterranean depths. Some characteristics are fixed within a species, while others can respond to environmental conditions.

In the same way that leaves compete for light aboveground, roots vie for available nutrients.[28] Nutrients are usually unevenly distributed in the soil.

For this reason, plants with roots that can grow directionally toward nutrients or that are especially efficient at acquiring or using resources will have a competitive advantage. Some nutrients exist in a form that roots cannot readily take up, so the competitively successful plant is the one that can access these nutrients. One method is to convert the nutrients into a soluble or transportable form. The root secretes compounds that increase the solubility of the nutrients or binds them so that they can be absorbed (see Chapter 3 for more on this topic).[29]

Another method is for the plant to recruit "friendly" microorganisms to convert the nutrients for the plant. But how does a plant ask another organism to take on this task? One way is to secrete fluids into the soil surrounding the roots in order to alter soil pH or micronutrient composition and thus attract bacteria or other microorganisms capable of collaborating to transform nutrients into a bioavailable form.[30]

In the face of limited resources, including nutrients, water, or available space, the roots of plants may compete. Roots can detect the presence of other roots or physical barriers in soil and respond accordingly by inhibiting the growth of lateral roots and root hairs, which can result in competitive exclusion—isolation or segregation of roots whereby nearby plants restrict growth to avoid entanglement or competition.[31] Root competition is less intense when resources are abundant in soils.

Researchers have identified a competitive process in roots that is similar to what occurs aboveground: plants adjust the responses of their roots according to whether their neighbors are kin or strangers. An experiment using a dune plant, the Great Lakes sea rocket (*Cakile edentula*), showed that plants growing next to their siblings had less root mass than those growing next to strangers. Because they did not need to compete with their relatives, they were able to allocate fewer resources to their roots.[32]

Plants forge collaborative relationships not only with other plants, but with groups ranging from fungi to bacteria to insects. They release compounds into the air to attract insects needed for pollination, or to repel insect predators. Belowground, substances exuded from roots aid in cooperative processes. These exudates allow plants to influence their rhizosphere—the area around the root system—and the organisms that inhabit it. They can attract microorganisms that help the plant access nutrients, and they also play a role in kin recognition. Experiments have shown that root exudates allow some plants to distinguish siblings from strangers.[33]

The volatile chemicals that plants release into the air act as signals. When a leaf or stem is bitten by a herbivore, molecules are released that travel to other organs on the same individual, as well as through the air to neighboring plants. Danger! The recipient of the signal mounts a preemptive chemical defense response or other protective behavior to ward off damage.[34] Plants also employ volatile signals in other plant–plant interactions. Parasitic plants, for example, are able to identify a host plant by means of volatile chemicals, apparently using similar chemical cues as those used by herbivores to locate and distinguish among plants.[35] These attractive cues appear to be constitutively produced by plants, perhaps as secondary metabolites or metabolic byproducts; however, the production of airborne chemicals used to signal danger is induced upon herbivory or damage.

Volatile compounds are also involved in indirect protective mechanisms. When the leaves of corn plants are attacked by a butterfly or moth larva, for example, the plant releases a chemical that attracts a parasitic wasp, a natural predator of the larva. The attracted wasps feed on the larvae and prevent them from damaging the corn plant.[36]

In addition to communicating with other plants and with potential predators and pollinators, plants also build collaborative symbiotic relationships with other organisms. Symbioses—interactions between two dissimilar organisms that benefit both of them—are critical for plant growth and survival. Many plant roots have long-term interactions with nitrogen-fixing bacteria; the plants gain access to a form of nitrogen that they can use, and the bacteria obtain sugars from the plant.

Other important symbiotic relationships are mycorrhizae: associations between a plant and a fungus, in which the fungus improves water uptake and nitrogen and phosphate acquisition for the plant, and the plant provides food, in the form of carbon compounds, to the fungus.[37] Mycorrhizae have a critical role in community building and communication. A single fungus can connect multiple plants underground, resulting in extended networks and communities maintained via plant roots. At the same time, each plant may have a unique set of relationships with a different complement of fungi. Mycorrhizae establish resource-sharing networks by allowing all of the plants they interconnect to share carbohydrates.[38] Mycorrhizal associations are critically important for plants to survive and thrive; up to 90 percent of vascular plants have some type of mycorrhizal association.[39] Additionally, plants whose roots are connected via mycorrhizae can signal to each other. Experiments with bean plants attacked by aphids demonstrated signaling via mycorrhizae to connected beans plants, such that interconnected neighbors were warned of the presence of potentially damaging aphids.[40]

Just as in other aspects of plant behavior, kin appear to receive beneficial treatment. Investigators have found that common ragweed plants growing near kin had larger mycorrhizal networks than did those growing near strangers. In fact, kin plant communities have more plant–fungus interactions, which are associated with advantages for the plants, including the nutritional benefit of higher nitrogen content in leaves.[41]

Plant roots form mycorrhizae quickly, even before the plant could initiate longer-term solutions, such as root proliferation or development.[42] Plants can also adjust these associations to respond to changing environmental conditions. When light levels are low and photosynthetic efficiency is reduced, the mycorrhizal associations may decline.[43] Plants with limited energy stores or diminished abilities to replenish energy are not able to engage in nonessential behaviors such as symbiotic relationships. Under conditions of extreme resource limitation, a plant must focus on self-support; it cannot engage in sharing carbon compounds with fungi in exchange for access to phosphorus.

Mutually beneficial symbiotic relationships often extend beyond two partners, such as plant–plant or plant–fungus. Researchers studying an acacia tree native to dry regions of Africa and the Middle East, for example, discovered both a mycorrhizal fungus and a bacterium living within the tree's roots. Under stressful conditions of high salinity, acacia seedlings grew much better when they were inoculated with both organisms.[44] A similar tripartite symbiosis involving soil bacteria and mycorrhizae enhances the growth of mung beans and other crops.[45] Such synergistic networks are sometimes visible but are often hidden from view. Diversity in these networks supports integrated growth, maintenance, and functioning of the entire system.

We have seen in this chapter that the relationships plants form with others—whether other plants, insects, fungi, or bacteria—can be either collaborative or competitive; neighbors may be either friends or foes. In a competitive situation, however, plants have a number of methods that allow them to avoid expending excessive energy in antagonistic behaviors. And if their neighbors are kin, they often forge beneficial relationships with them. Choosing to collaborate can yield successes, sustain life, and promote longevity.

By studying how plants interact with others, we can see the importance of establishing an ecosystem of support, collegiality, and community. I think about how my own professional network has been greatly enriched not only by engaging with those who are closely aligned with my disciplinary focus as a biochemist, but by expanding my circle to include those who have strengths in areas to which I aspired, including mentoring and leadership, when I first began to work in those areas.

My collaborators and I have found in our own work that when we try to apply this model to human relations, we run up against the predominant mindset that focuses on individual success models.[46] For example, it is not uncommon to find individuals from marginalized or first-generation backgrounds shut out of local networks of knowledge in educational or professional settings, which can derail the likelihood of achieving success.[47] We often talk about this as people not gaining access to the unofficial or unwritten rules that are passed by word of mouth from those in the know. But we can learn a great deal from the network-based relationships that plants form. They provide examples that we can apply to building and sustaining personal, professional, and learning collaborations (such as community gardens, community-based mentoring programs, and collaborative professional work), and they offer a salient demonstration of the power of diverse communities.

The cultivation of symbiotic relationships and interconnected communities of support and shared, or reciprocal, value, in which each individual involved in an exchange both gives something and receives something in return, provides opportunities for individual success and a larger productive community.

As plants can teach us, responses to the environment need not be individual; at times they are best initiated collaboratively, whether in relationships of two or three, or as part of a wide-ranging network. The most effective networks are established and maintained by means of robust systems of communication and diverse interactions with collaborators and potential competitors. Perhaps humans could learn from plants to consider a broader definition of kin. Frequently our definitions of kin don't stop at those individuals who are our genetic siblings, but functionally extend to individuals with whom we share a similar demographic, be that racial, ethnic, or socioeconomic. We extend beyond genetic relations, but only narrowly, when those whom we have included are no more or less genetically related to us than those we exclude.

To improve our outcomes, we can benefit from including partnerships that branch out beyond our current biases of whom we extend kinship to apart from those who are biological kin. This effort requires the hard work of first recognizing and then confronting our biases. Yet if we are successful in this regard, broadening who we see and engage as our kin could dramatically change our current environment and the potential for success and thriving for all.

No risk is more terrifying than that taken by the first root. A lucky root will eventually find water, but its first job is to anchor.

—HOPE JAHREN, *Lab Girl*

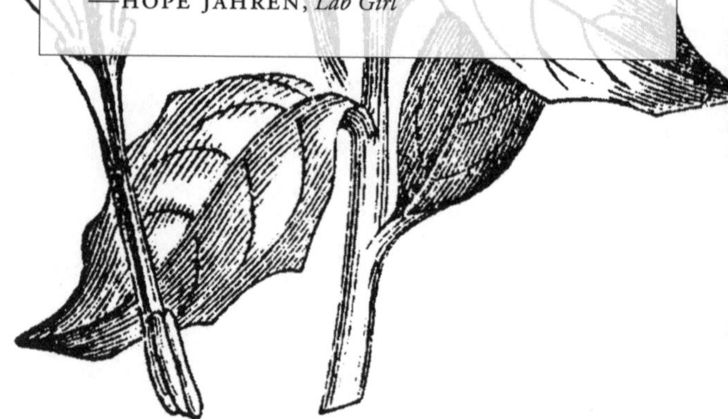

3

Risk to Win

You may have admired the yellow petals of canary creeper, the striking purple of prairie asters, and the bright orange flowers of marigolds along a roadside or in a field or garden. These are all different types of annual wildflowers—plants that complete their life cycle in a single growing season. If you are a gardener, you know to plant annuals (pansies, zinnias) every spring, but you count on perennials (daylilies, peonies) to come back year after year. In the wild, annuals typically emerge from seeds after a disturbance, like winter or a dry season. They grow, flower, and die during a short window of opportunity. To emerge in such uncertain times is a risky strategy, but it has advantages. These small plants limit allocation of energy to vegetative growth, choosing instead to grow fast and invest in flowering and seed development.[1] With a short life cycle, they avoid having to struggle with more robust plants for access to sunlight and soil resources. The risk of exposure to predators and herbivores must be weighed against the opportunity to have full access to sun and nutrients.

It is a decided risk for a seed to germinate. Should the seed germinate after only a splash of rain or a single warm day, or should it wait until the soil is thoroughly moist and temperatures consistently balmy? Some species have evolved to be risk-takers, with a low threshold for germination, while others, the risk-avoiders, wait for more reliable conditions.[2] The idea that plants are risk assessors might be new to you, but botanists have become aware that plants assess risks in much the same way that animals do. Such assessment underlies many of the activities they undertake. Some of these behaviors may be genetically fixed, while others are flexible and involve decisions taken during the plant's lifetime.

The way that plants perceive and weigh risks has potential to offer amazing insights for us. When local conditions are suboptimal, plants exhibit risk responses beyond what we might expect, especially when we are used to observing animals that have the ability to pick up and move elsewhere. The fact that plants live their entire life cycle in a single environment provides unique perspectives on effective risk-taking. Plants weigh risks and respond to scarcity in remarkable ways, all while staying put.

In the animal kingdom, decisions about whether to engage in risk-taking or risk-avoiding behavior are largely driven by environmental variation in resource availability and are influenced by concerns about energy use. Scientists have developed a concept called risk-sensitivity theory to predict how an animal responds to risk, primarily as related to its resource budget and strategic energy allocation. According to this theory, for example, an animal facing a predator decides whether to flee or defend itself by taking into account how much energy it needs for internal processes such as growth, activity, and reproduction, and for responding to external factors, such as temperature.[3]

For a long time we thought that plants did not engage in risk assessment, but many recent studies indicate that they do. Of course, plants behave in somewhat different ways from animals. Plants may respond to a threat by reallocating resources, whereas animals use the resources to flee (or fight).[4] But, like animals, plants tend to take more risks in dynamic or unpredictable environments and in times of scarcity. If a plant's roots are located between two environments, one with a constant but low level of nutrients and one with varying levels, the plant will choose to proliferate more roots into the area with varying levels. The plant is gambling on being exposed—even intermittently—to sufficient levels of nutrients.[5] This behavior is the same kind of risk-sensing that is found in animals. In conditions where there is a steady and adequate supply of resources, individuals take fewer chances. But when resources vary, individuals often commit to risky behaviors in order to increase the probability of long-term success.

Risk assessment and decision-making inform almost every stage of a plant's life cycle. From the time a seedling emerges, the plant is assessing its need for light and nutrients and making adaptations based on how available these resources are. Because plants are continually monitoring environmental cues, they can quickly sense when the situation changes and react with a short-term or long-term response, as appropriate. Scientists have found that plants are exquisitely sensitive to changes over space or time in the levels of resources. Remarkably, they can discern not only whether the concentration of a particular resource is changing, but how rapidly it is changing (that is, the steepness of the gradient).[6] Responding to such dynamic environmental conditions entails risk, but in the long term, such a strategy will improve growth and survival.[7]

Plants assess the potential return on investment for preferentially allocating energy to growth, reproduction, or defense depending on environmental conditions and resource availability. Volatile organic compounds (VOCs) can act as potent cues about present and future conditions, thus helping plants make decisions about how to allocate energy. As we saw in Chapter 2, plants that are being attacked by herbivores release VOCs that provide a warning signal to others. But should the plant receiving that signal devote more resources to protect against an attack that might not even occur? A fascinating study on sagebrush showed that plants that were provided with additional water were more likely to mount a defense in response to warning signals than were those receiving just rainwater—those with more resources available were more willing to allocate energy to defense.[8] In a similar experiment with pea plants, water-stressed plants communicated signals to unstressed neighbors, most likely transmitted through the roots, that served as a warning of danger. The neighboring plants responded with a stress response, possibly in anticipation of forthcoming drought.[9]

Risk-taking behavior is especially common when resources are variable or limited. Plants can respond by redistributing resources (short-term or long-term), finding ways to acquire more resources, ceasing growth, or, in the most extreme case, determining that the environment is unsuitable for continued existence. A flowering plant that finds itself lacking sufficient sunlight or nutrients for survival, for example, may shift its energy pools to the production of seeds. These seeds might be transported by wind or animals to a different environment, or they may fall to the ground and be stored until conditions are more favorable.[10]

Plants act as "dynamic strategists," changing their behavior based on their perception of stress or environmental limitations.[11] Let's explore this first in the case of nutrients. Plants that have access to constant, high levels of nutrients do not need to take risks. They simply distribute their roots in the nutrient-rich space.[12] When the nutrient supply is low or patchy, it is risky to initiate an energy-requiring process. Yet, some plants do use energy to stimulate root proliferation and elongation in these situations, because the benefit of encountering scarce nutrients outweighs the cost of producing new or longer roots.

Other responses to low nutrient availability include breaking down chlorophyll (degreening) to reduce cellular metabolism that is dependent on the limiting nutrient, or increasing the ability to take up target nutrients from the soil.[13] Notably, plants with limited resources also happen to be much more precise in their resource allocation decisions—perhaps because the risks of reduced nutrient uptake and detrimental effects on growth and reproduction are greater if a wrong decision is made.[14]

Iron is a key nutrient for plants, as it is essential for photosynthesis. It is contained in the photosystems that absorb light, and it functions in the cofactors needed for the chemical reactions of light harvesting.[15] The iron that is present in soils, however, is often in an insoluble, oxidized form—equivalent to rust—that cannot be taken up by plant roots or used to synthesize compounds that support metabolism and photosynthesis.[16] Plants engage in several different strategies to solve this problem, depending on whether the iron limitation is slight or severe.

Some plants can increase their iron uptake through the use of chemical compounds called siderophores, which bind and transport iron. This strategy is used most commonly by grasses.[17] The siderophores are excreted through the roots into the soil, where they create complexes with iron. The iron-siderophore complexes are taken up by means of specialized proteins called transporters.[18] The plant cells then convert the iron from an insoluble to a soluble form, which is released for metabolic use.

Other plants, mostly non-graminaceous monocotyledons (that is, herbaceous plants apart from grasses) and dicotyledons, use different strategies to acquire iron.[19] One of these involves excreting protons from the roots, which increases soil acidity and makes the iron more soluble. Another depends on engaging with particular soil microbes that synthesize their own siderophores.[20]

Other nutrients besides iron are essential for plant physiology, structure, and function. Nitrogen plays a critical role as a component of amino acids (the building blocks of proteins) and of chlorophyll.[21] As in the case of iron, short-term nitrogen limitation prompts responses that increase nitrogen uptake and utilization. Some of these responses involve structural or developmental changes, such as changes in the morphology of roots.[22] Plants initiate proliferation of the root system to increase foraging for nitrogen unless the deficiency persists for an extended period. At that point the plant may restrict root development to conserve energy for survival or reproduction.[23] Root proliferation entails a significant energy investment and can be risky, since the plant is gambling that investing in a more extensive root system will increase the chances of encountering a high-nitrogen patch.

And as we've seen with other kinds of strategies intended to gain access to resources, plants can choose either individual or collaborative responses. This is the case with nitrogen availability, as well. Many plants respond to limited nitrogen availability by forming synergistic relationships with nitrogen-fixing bacteria. These bacteria may be located inside the root, in structures known as nodules, or on the surface of the root.[24] This symbiotic interaction involves a bilateral exchange that is beneficial to both partners. The plants transfer carbon to the bacteria, and the bacteria produce nitrogen in a form that can be readily taken up by the plant.

Another important nutrient for plants is phosphorus, which is naturally present at relatively low levels in soils.[25] Phosphorus is critical for development, growth, and maintenance, as it is a component of the nucleic acids DNA and RNA, as well as the energy storage molecule ATP and the phospholipids that are present in cell membranes.[26] When faced with a phosphorus deficiency, plants pursue several different strategies. One option involves increasing phosphorus solubility by altering soil acidity through excretion of protons, similar to what happens in iron-deficient conditions.[27] In a longer-term adaptation, more energy may be directed to root proliferation. This kind of response is similar to what is observed under nitrogen limitation.[28]

As in the case of nitrogen limitation, one long-term solution for coping with low levels of phosphorus is collaboration. Some plants, as we saw in Chapter 2, have evolved the capacity to interact with fungi by forming mycorrhizae. This symbiotic partnership enables the plant to absorb phosphorus from the soil more efficiently.[29]

It is important to keep in mind that when plants engage in symbiotic relationships to increase access to nutrients, they are still taking risks. In committing energy to forging such relationships they are trusting in reciprocity, as well as trusting that better access to resources will increase their fitness and persistence. Thus, the plant anticipates that the payoff in working together to increase resource acquisition will be greater than the cost of producing sugars that are gifted to their fungal partners. That is not always the case. Under some conditions, carbon costs to the plant outweigh the nutrient benefits received in return, which shifts the mycorrhizal association from symbiotic toward parasitic.[30]

In the same way that plants change their behavior based on nutrient availability and assessment of risk, they must also take into account the availability of other vital, variable environmental factors, especially light and water.

When a plant does not have access to adequate light, whether due to shading or competition, it must adapt. One long-term structural adaptation to limited light availability is to change the architecture of the leaf. Leaves growing in full sunlight, known as sun leaves, are thick and have more palisade cells than spongy mesophyll cells. The palisade cells contain a large number of chloroplasts—the engines of photosynthesis—while the spongy mesophyll cells, which make up the inner tissue of a leaf, contain fewer chloroplasts and have more intercellular space. Compared with sun leaves, shade leaves are thinner; they have less chlorophyll, and more spongy mesophyll cells than palisade cells.[31]

Constructing a leaf is a costly—and thus risky—investment. A particular leaf structure optimizes light capture and the conversion of light to chemical energy in a specific environment. The same leaf will be less successful in a different environment. Sun leaves do not function well in shade, as they possess too much chlorophyll for the available light. And shade leaves are vulnerable to phototoxicity in full sun: they produce lower amounts of photoprotective pigments as a tradeoff for investing energy in a distinct leaf architecture and other shade-relevant physiologies.[32] Thus, investments in leaf architecture carry risks, because the leaf may end up in a light environment that does not match its form. Plants weigh these risks by assessing whether exposure to a particular environment or to a particular level of a resource (abundance or limitation) is likely to be short- or long-term. If it appears to be long-term, then taking the risk to alter leaf morphology may be desirable.

There are also risks associated with changing other aspects of plant architecture, such as growing new shoots or branches. Plants can regulate the number and size of shoot branches, the initiation and development of which is an energetically costly process, to manage environmental risks. In some situations it pays to invest in additional branches and leaves, which can support more flowering and seed production. But in other situations, it may be preferable to limit growth and flower quickly, before environmental conditions deteriorate. Researchers studying Mediterranean annuals found that the plants assessed the risks and potential costs of investing in large vegetative structures based on the reliability of environmental cues. They adjusted their growth patterns more in response to reliable cues, such as day length, than unreliable cues, such as water availability.[33]

Another risk-associated behavior is related to water conservation. Leaves have small pores, called stomata, that take in carbon dioxide and expel water vapor. Plants adjust the opening and closing of stomata to regulate water balance, and they have evolved different strategies based on considerations of risk. Botanists divide plants into two broad categories according to how they regulate water status. One group, called isohydric plants, maintain a relatively constant water content in their leaves. They do this by closing their stomata during dry conditions to keep water vapor from escaping. Although this strategy preserves water, it has the disadvantage of reducing the amount of carbon dioxide taken in, which lowers the rate of photosynthesis and the production of carbohydrate compounds used for energy. The other group, anisohydric plants, do not maintain constant water content in their leaves. Under dry conditions they keep their stomata open longer, thus maintaining higher rates of photosynthesis. Keeping stomata open is risky because the plant can dry out too much. If the plant survives, however, it will likely have a fitness advantage over a plant that chooses to conserve water, because it has been able to maintain photosynthetic productivity.[34]

As we have seen, plants are constantly taking risks when considering opportunities and making decisions about where to invest their energy. Plants that invest poorly may not survive, while those that make good decisions will thrive.

The prairie aster growing by the roadside, like all plants, must assess risks in responding to immediate environmental conditions, but its entire life strategy is something of a gamble. This plant has evolved a different life history from one that lives many years. Annual wildflowers put all of their energy into growing—and growing quickly—during the literal sunlight "window" of opportunity. Because their lives are short, they have a better chance of avoiding predators than a longer-lived plant. If they survive and reproduce, they will leave seeds stored safely in the ground, ready to emerge the next growing season, or after a disturbance. In the meantime, larger, perennial competitors will just be starting to emerge and establish their dominance in the ecosystem. In the long run, the annuals' risk pays off.

The risk-taking and risk-avoiding behaviors that plants engage in reveal wise ways of being that we humans would do well to emulate. They rely on careful environmental sensing to provide information that allows them to identify potential risks and guide decision-making. They assess what resources are in short supply, what collaborators are available to help alleviate specific resource limitations, and how to initiate and maintain collaborative relationships for improved resource acquisition. They decide where to allocate energy depending on what risks they can take. To survive and thrive, they must consistently scan and assess all aspects of their surroundings, including availability of light, water, and nutrients, as well as the plants, bacteria, fungi, and other organisms that are near them.

We humans could learn how to better sense our surroundings, assess risks, and support one another in the way that plants do. We should support each other's short- and long-term goals, opportunities, decisions on how to allocate or redistribute our resources, and appropriate timing of personal or professional transitions tuned to environmental parameters, whether the goal is individual or community growth. To accomplish these tasks we must be adept at environmental surveillance. Since we have a finite amount of energy available in a specific period for all of our activities, we must take care in deciding where to allocate energy and which risks are worth taking. Just like plants, humans need to make strategic decisions about how to use our finite energy resources to maximize growth and success in our dynamic environment.

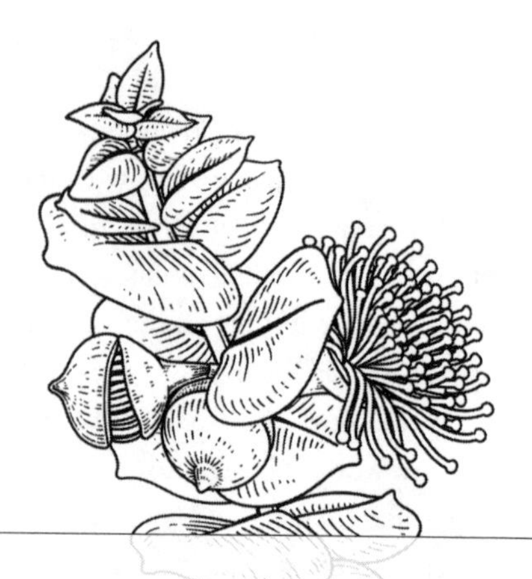

Desire makes plants very brave, so they can find what they desire; and very tender, so they can feel what they find.

—AMY LEACH, *Things That Are*

4

Transformation

I drive by an abandoned factory lot on my way to work daily. Year after year, I have watched it change. At first it was barren, then it became grass covered, and now it harbors a community of flowering plants, small bushes, and young trees. It has been absolutely fascinating to observe this desolate area develop into a successful mixed community of plants. I have been a witness to plants progressively transforming the land into a rich ecosystem.

The progression that has occurred on this urban lot is similar to what occurs in the wild after a disaster like a volcanic eruption or a flood. After an eruption, the lava flows down the mountainside, scorching and destroying everything in its path, and covers the land, hardening as it cools. The result is the construction of a new habitat: nearly sterilized land that is largely devoid of living organisms. This was the scene that took place on Mount St. Helens in 1980, when a volcanic eruption was followed by a landslide caused by partial collapse of the mountain. The event left complete destruction and massive land clearing in its wake. Eventually, plants began to grow back. Some seeds had remained in the soil and were able to germinate. Others were brought in to the area, deposited by birds or wind currents. And some plants regenerated from roots or branches that had survived the eruption.[1] After a disturbance like this, the rate at which plants become established is determined by the amount of moisture available, as well as by the ability of the colonizing plants to take root and survive on the limited nutrients present in volcanic ash or hardened lava.[2]

We have seen how capable plants are at thriving no matter where they happen to be situated, because of their impressive abilities to perceive what is going on around them, to adapt, and even to alter themselves or their environment to better support their growth and persistence. In this chapter we focus on their ability to transform their environment to make the resources they need more readily available.

A volcanic eruption is just one example of a disturbance that transforms plant ecosystems. Another example concerns fire. After a fire rages over the land, disrupting the ecosystem, the soil may be left largely barren and subject to erosion, or some organic matter or vegetation may remain.[3] Plants eventually return, sometimes rather rapidly. Many factors determine which species persist or emerge after the fire, including the fire intensity, the pre-burn species composition, and the makeup of the seed bank. The germination of some seeds, such as those of many pines, eucalyptus, sequoia, aspen, and birch, is activated by fire or smoke.[4] Other plants, like many grasses, some oaks, and eucalyptus can regenerate from roots post-fire.[5]

Plants have even been able to reestablish themselves in highly toxic environments, such as Chernobyl, Ukraine, the site of a catastrophic nuclear radiation disaster in 1986.[6] Many coniferous trees, such as Scots pines, died after the nuclear event because they are very susceptible to radiation. The land regenerated rather quickly, as deciduous trees that are more resistant to radiation grew back.[7] Not all of the trees in the area died, however, and those that survived provided valuable study material. To measure the impact of the radiation catastrophe on the growth and recovery of trees before and after the event, scientists took cores from them. They assessed the cores for ring width, which is a good proxy for radial growth, as well as wood quality.[8]

Tree rings are formed by a thin layer of cells called the vascular cambium, from which the water-conducting tissue known as xylem is formed. Xylem constitutes what we call wood.[9] Each tree ring represents a year's growth of xylem, and its width represents its relative annual growth. Comparing the widths of the rings of a core that extends from the outside of the tree to the center yields insight into season-to-season variations in growth. Properties of the wood, such as porosity, can also be studied with cores.[10] Researchers looking at the cores from Chernobyl trees discovered that the higher the level of radioactivity to which they had been exposed, the more slowly they grew. They grew the most slowly right after the disaster, when radiation exposure was highest. Radiation-exposed trees also exhibited long-term impacts on growth and wood composition that were apparent for up to ten years after the accident.[11]

In addition to directly damaging trees, the radiation extended into other parts of the ecosystem that affect the trees' establishment, growth, and persistence. For example, it led to the loss of many invertebrates and bacteria that live in the soil and decompose the leaf litter and other organic debris that collect on the forest floor. This disruption in the natural processes of replenishing and maintaining soil health led to significant changes in the soil ecosystem and inhibited the growth of many species of plants.[12]

Because plants are resilient and tend to recover from disasters more quickly than animals, they are essential to the revitalization of damaged environments. Why do plants have this preferential ability to recover from disaster? It is largely because, unlike animals, they can generate new organs and tissues throughout their life cycle. This ability is due to the activity of plant meristems—regions of undifferentiated tissue in roots and shoots that can, in response to specific cues, differentiate into new tissues and organs. If meristems are not damaged during disasters, plants can recover and ultimately transform the destroyed or barren environment. You can see this phenomenon on a smaller scale when a tree struck by lightning forms new branches that grow from the old scar. In addition to regeneration or resprouting of plants, disrupted areas can also recover through reseeding.

Scientists studying the vegetation growing around Chernobyl have found an additional protective response that has reduced the damaging effects of radiation. Radiation causes harmful genetic mutations in all organisms, but plants that were exposed to it for many years developed adaptations that helped to stabilize their genomes.[13] This is yet another powerful illustration of how resilient plants are, and of their abilities to persist in and potentially transform environments. This ability to remain resilient in the face of environmental challenges, as well as to transform the environment through one's persistence and continued growth and thriving, are important characteristics that humans would do well to adopt.

After a volcano, fire, or other disaster, as plants, animals, and microorganisms return to an environment, the composition and structure of the ecosystem change in often predictable ways. Grasses may give way to shrubs and then trees, for example. Ecologists use the term "succession" to refer to these long-term changes, and they distinguish two different kinds: primary and secondary succession. Primary succession occurs on new land or rock formations where no soil is present, such as on hardened lava flows or on islands that are newly emerged from the sea. Secondary succession refers to the establishment of a community or ecosystem after a less severe disturbance, such as a fire or flood that has not removed all vegetation and soil.[14]

As you might expect, numerous factors influence which species first emerge after a disturbance and how the species composition changes with time. Some of these include the availability of nutrients and of light—both the amount of light and its spectral properties. By doing experiments that involve removing and adding different plant species, biologists have found that succession patterns are strongly influenced by the particular species that are present in an area and the interactions that occur among them.[15] Other factors that influence succession are climate change and the presence of invasive species—species that are not native to a particular ecosystem and that cause some form of ecological or economic harm.[16]

As succession proceeds, the structure of the community changes, as do other aspects of the ecosystem, such as soil characteristics.[17] The abilities of individual species of plants to adapt to the local environment through the processes of colonization, establishment, and growth and survival significantly affect the overall pattern of succession.[18]

A number of key attributes determine which plant species thrive following recurrent disturbances such as fires. These attributes include the method of persistence through the disturbance, mechanisms of establishment, and the time taken to reach critical life stages (reproduction and senescence). Persistence has to do with whether a species has attributes that enable it to return after the disturbance; it may have seeds that remain viable in the soil, or it may be able to sprout from surviving roots. Mechanisms of establishment involve how plants are able to grow and thrive after a disturbance. Some may be able to establish themselves quickly while others may come in later, depending on factors such as the ability to compete for resources. The time it takes to reach critical life stages is also crucial; the time needed to mature and reproduce, for example, affects how quickly a species can establish dominance.[19]

There is a great deal of variation in patterns of succession, but scientists have suggested that they fall into three distinct pathways—namely, facilitation, tolerance, and inhibition. These pathways describe whether species that become established early in succession facilitate, tolerate, or inhibit the establishment of later species. The facilitation pathway is more common in primary succession, whereas the tolerance pathway often characterizes secondary succession, where soils and nutrients may be readily available. Inhibition occurs when established species inhibit invasion by competitive species. This inhibitory state persists until established species senesce or suffer considerable damage, in either case resulting in resources being made available for other species to colonize a niche.[20]

While interplant competition certainly plays an important role in succession, interactions with other organisms are also influential—these include animal grazing or other forms of herbivory, as well as the presence of pathogens.[21] Herbivores can limit a plant's growth and production of seeds, and thus limit potential for plant dispersal and persistence.[22] They can also influence nitrogen dynamics and chemical properties of soils, and related feedback effects on the life cycle and persistence of plants and plant communities. Herbivores can lead to reduced cycling of nutrients such as nitrogen in an ecosystem by reducing plant biomass with nitrogen-rich tissue, which they preferentially consume.[23] Herbivores can, thus, cause regression in the rates of recovery or shift species dynamics during succession.

In primary succession, the first plants to emerge in a barren environment are called pioneers. These species are able to grow despite significant environmental challenges. Pioneer plants are those that you might see sprouting from a crack in the sidewalk or a driveway. They can also be found emerging from hardened lava. These plants have the ability to track trace moisture in the most minute of cracks, allowing them to grow on the edge of a cliff or in crumbling asphalt, where there is rare access to moisture. As a life history strategy, the limited—perhaps once in a lifetime—opportunity to access critical moisture or sunlight can outweigh other risks of growing in such spaces.

Pioneer plants generally have minimal requirements for resources and are good scavengers. They can grow in different types of soil and are able to deal with very low nutrient availability. Indeed, many pioneer plants are able to increase nutrient availability—either through excreting compounds that increase the solubility of certain nutrients such

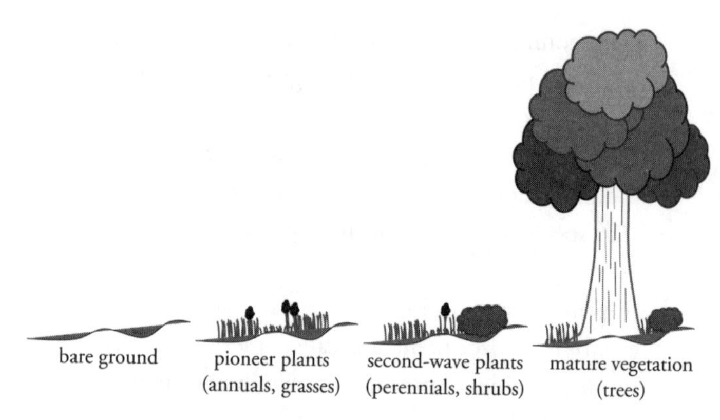

Primary succession begins with a barren environment that has experienced a severe disruption such as a fire, flood, or volcanic eruption. Pioneer plants emerge first; they require limited resources and can root in poor soils. Pioneer plants improve soil properties, allowing the establishment of a second wave of species that have higher resource requirements. Ultimately, with continued improvements and changes to the ecosystem, trees emerge, along with other plants that require rich soils and can grow in shade.

as iron, or through engaging in relationships with other organisms, such as nitrogen-fixing bacteria or fungi that form mycorrhizae.[24] Pioneers produce several effects that improve conditions for later arrivals; they change the pH of the soil, making it more favorable for other plants, and their presence increases soil stability and reduces the impact of damaging winds.[25]

As pioneer plants grow, they transform the environment in ways that make additional resources available, including providing access to soil that may lie beneath sidewalks, pavement, or lava, or loosening highly compacted soil. Each individual creates a new microclimate—a local climate that may differ from that of the larger ecosystem. The created microclimate can support the plant's own growth, as well as contribute to the success of later-emerging species that have greater demands than the earlier-emerging hardy plants.[26] Some pioneer plants are able to break down rocks or lava through the mechanical force exerted by growing and expanding roots or by excreting acids or other erosive chemicals from their roots.[27]

These first emergers are usually adapted to grow in dry soils in areas with exposure to bright light.[28] After they die and decompose, they contribute to the formation and enrichment of the soil.[29] Through this and other means, minerals and nutrients in soils typically increase with time, but this transformation occurs slowly during primary succession. Limited resource availability may continue to restrain community growth and development.[30] The pace of succession is also affected by the properties of the species that follow the pioneers.

Plants that emerge in the second wave of primary succession have slightly higher nutrient requirements. Yet, these plants can generally still grow in soils that are not considered particularly rich in nutrients. Like pioneers, these plants are often adept at acquiring scarce nutrients, or they collaborate with organisms that support the transformation of resources into more accessible states. As the activities of these second emergers transform the environment into one with more abundant resources and accessible soil patches, additional waves of plants can begin to flourish—those requiring more nutrients and fertile soil or that can grow in shaded or reduced light environments. The sequential establishment and success of different plant species ultimately leads to a more diverse ecosystem, although diversity may be highest earlier in succession, as dominant species become stable and can inhibit entry of additional species later in the process.[31] Some of the plant properties that affect the order of succession in disturbed environments include those having to do with seedling establishment; the ability of a seedling to germinate and take root successfully can vary based on evolutionary history and local ecological conditions.[32]

Flooding is a different kind of disturbance from a volcanic eruption. While flooding can result in widespread disruption of an ecosystem, it does not completely destroy it. A massive flood, such as that experienced during a hurricane, often kills many small plants outright and deposits soil or silt that buries others. Larger plants and trees frequently will survive but may suffer severe physical damage. After an ecosystem is disrupted by flooding or by wind, fire, or another such disturbance that causes severe damage, secondary succession takes place. The area, which has not been completely stripped of all plants and other living organisms, is subsequently recolonized or reinhabited.

The pioneer species in the case of secondary succession often possess somewhat different qualities from those that thrive during primary succession because more nutrients are available, as well as access to soil. There tends to be less competition for resources during secondary succession than primary succession.[33] A principle of ecology is that no two species can occupy the same niche (that is, play the same ecological role) in the same location; one will outcompete the other. The process by which one species replaces another during succession occurs at varying rates and affects the ultimate species diversity in a community.

Species diversity can be measured in several different ways. Ecologists often refer to alpha diversity, defined as the number of species found in a local area, and beta diversity, or the variation in species composition between different areas in a region.[34] Diversity in a local area is affected by many factors, including how easily different species are able to move between patches.[35] Alpha diversity usually increases during the course of succession, although in some ecosystems, environmental stability develops over time and diversity can then decrease.[36]

So far we have been talking about processes in natural environments. In urban landscapes, and other land that has been altered by humans, diversity follows different patterns. Human interference is an important factor in altering local diversity and affecting plant demographics. For example, vacant lots with limited human interventions have a large number of different species of plants, but generally the same plants are found in every lot (that is, high alpha, but low beta diversity); by contrast, residential gardens possess fewer species in each garden, but lots of variation between gardens (representing low alpha, but high beta diversity).[37]

Roots play an important role in succession because of their influence on plant establishment and their transformative properties. Underground, just below our feet, roots are exerting control on soil properties and, thus, on entire ecosystems. A plant's health is determined in large part by the activity and function of its roots. We can gauge a plant's health by its ability to form blossoms and fruit, but it is the roots that provide the necessary nutrients for reproduction. Plants acquire nutrients from the soil, and, as we saw in Chapter 3, when nutrients are scarce, they gain access to them by altering the morphology of the roots—shape, length, branching—or by exuding compounds to increase the solubility of the nutrients. These actions can transform soil quality and promote collaborative interactions with bacteria and fungi.

Among the most dynamic parts of terrestrial ecosystems are those related to roots, including the layer of soil that adheres to the root hairs, the rhizosheath, and the soil that surrounds the roots, the rhizosphere.[38] Activities associated with these soil elements drive many aspects of plant establishment, persistence, and transformative ability. Both the physical makeup of roots, as well as compounds that are produced by roots, impact rhizosheath production and rhizosphere properties and function. Whether soils are compact versus loose, or nutrient poor versus nutrient rich, directly affects seed establishment and plant longevity.[39] These root responses transform soil characteristics directly and can in turn affect the physiology and ecology of all soil inhabitants. This behavior is indeed transformative.

Roots' plastic properties—their ability to change in response to environmental conditions—can be either biochemical, as in the production of exudates, or physical, involving structural alterations. Plant roots release solutes and carbon and can exhibit structural differences in response to cues from the environment, resulting in changes in the soil ecosystem. For example, root structure can influence water dynamics in the soil.[40] Changes in root architecture and biomass modify soil porosity, likely by altering soil compaction—which can in turn alter how water is absorbed and flows through soil, ultimately affecting plant responses such as water uptake. Some of these responses, including root exudation, can be controlled temporally, which enables the plant to adapt in ways that are both rapid and reversible.[41] Other responses, such as changes in root architecture, are long-term and result in enduring changes in the soil and the entire ecosystem.

Much of the dynamism in the ecosystem related to root function is due to exudate production and release by plant roots, as well as by associated microbes.[42] Root exudates can change the solubility of minerals and nutrients in soils—soil chemical properties—and can even detoxify detrimental materials, such as aluminum.[43] One substance excreted by roots that has a significant impact on the rhizosphere is mucilage, a gelatinous type of solution that contains sugars and glycol- and phospholipids.[44] Mucilage likely contributes to the drought tolerance of some plants. It can increase or significantly alter the ability of roots to transport water to the xylem, the retention of water by the rhizosphere, and the uptake of water relative to surrounding soil that lacks mucilage.[45]

Plants also synthesize and excrete lipid-based compounds that can serve as surfactants (wetting agents or dispersants). These compounds increase availability of resources for root uptake and plant utilization. Surfactants have been found to increase the solubility of phosphorus- and nitrogen-based compounds.[46] The change in soil properties, including increased availability of resources, also influences microbial physiology and processes, which additionally transform the soil biochemical and biophysical properties that support plant growth.

Mucilage and surfactant production and function are powerful examples of how products exuded by plants can transform soil habitats. But plants are not the only organisms affecting the composition of plant communities, especially underground. Fungi produce similar compounds called sterols (which are related to cholesterol) that are water-repelling and keep fungal hyphae from drying out, thus increasing water retention of the rhizosphere.[47] These organisms also release hydrophopic glycoproteins that coat soil aggregates, thereby modifying the soil's capacity for absorbing water.[48] Sterols, glycoproteins, and other fungal products affect the biochemical and biophysical properties of soils in the same way as the mucilage produced by roots.

Soil composition and soil microbes also have significant effects on ecological succession and environmental transformation.[49] These mostly unseen portions of ecosystems—and the myriad interactions occurring among them—change over time and have corresponding long-term effects on the plant species composition of a community, as well as on ecosystem development.[50] The composition of fungi in soils also changes as succession proceeds, as do the nature and function of mycorrhizae, which in turn influences plant community composition.[51]

The complex interactions of mycorrhizae with plants and soils, although hidden from view, drive critical aspects of succession and ecosystem change.[52] Recall that mycorrhizae are symbiotic relationships between fungi and plant roots that increase the roots' uptake of water and nutrients. Mycorrhizal fungal type and presence, together with soil fertility, affect what plants grow in an area because the effects of mycorrhizae on plant growth depend on soil quality for certain species. Some plants grow better in relationship with fungi only when in poor soils; when they are in rich soils, they do not show much benefit of mycorrhizal associations.[53]

Mycorrhizal fungal species also affect the competitive abilities of plants, perhaps by contributing to nutrient and mineral uptake and use. Scientists have shown that environmental conditions that limit a plant's ability to sustain high levels of photosynthesis, such as poor soils or shade, can also limit its abilities to form mycorrhizae.[54] These plants would have a competitive disadvantage compared with those that had better-developed mycorrhizae, which could affect population composition and dynamics over time. Plant species composition, in turn, can change the soil populations of mycorrhizal fungi.[55] As shifts in the soil fungal population occur, the range of plants that can be supported in the soil may also shift. These changes could, thus, result in soils that have the potential to support only specific plants with mycorrhizal requirements that match the fungi present in the soils.[56] In other words, the impact of plants on mycorrhizal fungi population structure and dynamics can influence plant succession, and thereby the composition of current and future plant communities.

Another example of collaborative behavior among plants, with the potential to transform the environment, is a phenomenon called swarming. Swarming is a form of social behavior that is based on mutual engagement between distinct individuals, and it can serve as an emergent strategy, enabling complex patterns to be built up through small interactions.[57] It occurs when a number of individuals all move together in the same general direction in either an active or passive way—active swarms are self-generated rather than produced by outside forces.[58] Everyone is familiar with some everyday examples of swarming behavior, like flocks of birds, schools of fish, or, of course, swarms of insects. Swarming behavior is also common in bacteria. Swarming bacteria have been shown to move toward areas with rich pools of nutrients and limited competition for resources.[59] Adrienne Maree Brown describes such behavior eloquently: "There is an art to flocking: staying separate enough not to crowd each other, aligned enough to maintain a shared direction, and cohesive enough to always move towards each other. (Responding to destiny together.)" [60] Certainly there are clear lessons for humans in the importance of finding individuals with whom one can pursue personal goals in community with those moving in a shared direction and pursuing aligned purpose.

No one expected to find swarming behavior in plants, since they can't move around. But some plant parts do indeed move, and in 2012, a group of scientists announced that they had discovered that growing plant roots engage in active swarming. They found that the roots of neighboring corn seedlings tended to all grow in the same direction, even though they were in a homogeneous medium.[61] The purpose of such behavior may be to "optimize interaction with their environment."[62] One potential advantage of root swarming would be that a group of cooperating roots could release compounds such as siderophores to improve nutrient solubility in soils locally.[63] Swarming behaviors such as this would result in spatial regulation of soil chemistry and promote plant growth and endurance. Like the flocking of birds, the swarming of roots is an emergent strategy of shared destiny, yet it contributes to transforming the environment when the roots work together to solubilize nutrients or engage with other organisms such as bacteria or fungi symbiotically.[64]

"Bloom where you're planted." This phrase is often used to encourage people to survive and thrive wherever they find themselves. The idea is that we should behave like plants, which are widely assumed to make the best of the spot where the gardener places them. However, the analogy is misleading. As we have seen in this chapter, plants don't just function within their environment: they actively participate in and transform it. They mount phenotypically plastic responses to optimize their growth, and they demonstrate a sort of awareness that extends beyond the boundaries of their own selves and reflects knowledge of the external environment—what is sometimes called "extended cognition."[65] This awareness can lead to behaviors and adaptations that transform the environment, improving circumstances for the individual itself and for other inhabitants. In the process of succession, the early emergers affect the ecosystem in ways that determine which species will be able to grow and thrive in the next stage.

Promoting change in human environments requires similar skills to those that plants display during ecological succession. In human institutions, or ecosystems, effective initial leaders of cultural change function as pioneers. Identifying and supporting individuals who possess the characteristics needed to promote change successively and synergistically toward developing and sustaining a new ecosystem is critical. Leaders who are effective trailblazers, like pioneer plants, are able to thrive while guiding change with limited or variable resources. They also recognize that even when an environment seems stable, the efforts of the pioneers can forge new directions and innovation.

In human succession patterns, organizations often focus on group dynamics rather than recognizing and embracing the impact that individuals—especially effective change agents—can have on influencing desired cultural change. Accomplishing change requires leaders and trailblazers capable of pushing through obstacles, much as pioneer plants in primary succession may need to emerge through barriers or put down roots in difficult places. These pioneer individuals can often effect change with minimal resources or networks to support ideas, growth, and innovation. The effort of these individuals then leads to additional ecosystem changes that support the next wave of individuals needed to drive and sustain cultural change and institutional transformation.

The transformational goals of pioneers often require an initial period of disruption. Just as prescribed fires are needed for managing certain ecosystems, intentional disruptions may be required in human ecosystems to interrupt entrenched patterns, or status quo thinking or action, and to move purposefully toward intended change outcomes.[66] Although intentional disruption is often necessary, we should not overlook the reality that beneficial disruption opportunities can arise from bad intentions. For example, the 2016 election of a US president whom many Americans viewed as anti-woman and anti-science led to a national protest movement, including the 2017 Women's March and the March for Science.

Disturbances in an environment can change the composition of the individuals that are able to exist, thrive, and persist there. Yet, we tend to ignore the need to introduce intentional disturbance or disruption, just as we do in fire-adapted ecosystems; significant changes to the composition of individuals may be required to move toward change targets. People often purport to desire significant changes to ecosystem structures in the pursuit of equity but ignore the need for real "disturbance" to break away from the status quo community composition. An organization may need to reevaluate recruitment strategies and screening processes to promote the identification and recruitment of a broader range of individuals. We must understand that intervention and intentional disruption may be critical for supporting environments primed for the succession needed to support cultural change.

We can accomplish systemic change through multiple means, much as plants do. Strategic ways to initiate transformation start with reflection on and awareness of our current situation: identify the features of the local surroundings and what resources are available, and assess our needs. At the community level, a subset of individuals, pioneers, can serve the whole by acting as "sensors." Such individuals are positioned to assess climate or rapid changes in the environment which require responses or innovations. Intentional interventions in human ecosystems designed to support catalyzing a platform for long-term evolution of ecosystems can lead to desired outcomes. Everyone must recognize the importance of pioneers—those that have the traits needed to initiate robust culture change at the right time and place—and support the need for leaders to function in this pioneering way.

Nature's balance was created by diversity which might in turn be taken as a blueprint for political and moral truth.

—ANDREA WULF, *The Invention of Nature* (paraphrasing Alexander von Humboldt)

5

A Diverse Community

In the summer, I often visit a field filled with wildflowers. Some are so small you almost overlook them, while some reach a foot high and others stretch twice that, and they sport a rainbow of blooms. I am captivated by the variety of forms and colors that make up the community. As I gaze at the spectacle, I contemplate how all of these different species manage to coexist. Although I am filled with awe by the diversity in this place, many people walk, bicycle, and drive by without noticing this thriving ecosystem—they are lacking in plant awareness. I ponder how they could pass by without pausing, if only for a moment, to marvel at the array of plants in the field and consider the complex interactions that are going on both above ground and under the soil.

Scientists studying biodiversity in plant communities have found that many different species can peacefully coexist in part because of a phenomenon called niche complementarity. Each species occupies a slightly different niche—a position in the community defined by its life history, use of resources, and interactions with other species. Because each species, and even each genetic variant within a species, has different needs, the result is maximal use of resources in a particular community or ecosystem.[1] Not only does diversity benefit individual plants, but the unique abilities and behaviors of different species benefit the collective. Ecosystems with greater biodiversity tend to be more productive; that is, they produce more biomass—more leaves, stems, fruits, and other plant parts.

Today, commercial agriculture is characterized by vast monocultures of corn, soybeans, and wheat. While this practice makes planting and harvesting easier, it is not the only way to grow crops. Farmers in Indigenous cultures around the world have long used a technique called intercropping, which involves planting two or more crops together. Just as in natural ecosystems, it turns out that productivity is higher when certain crops are planted together in polycultures, rather than in monocultures.[2] Intercropping increases the productivity of individual plants through a process known as interspecific facilitation. Each species contributes something that promotes the growth, reproduction, or persistence of the others.[3] Because individuals in each species use different strategies for acquiring resources, they are able to divide up the resources instead of competing for them.

One of the best examples of intercropping is a venerable planting style called Three Sisters. This method of cultivation, which involves planting corn, beans, and squash together, has long been practiced by many Native American peoples.[4] With deep respect for the gift of the Three Sisters and other traditional ecological knowledge, but with no intention of co-opting that knowledge, I explore in this chapter what wisdom can be gained from deep contemplation of and reflection on this practice.

Why is the Three Sisters system so widespread? By planting corn, beans, and squash together, the planter is able to draw on their complementary strengths. Corn provides vertical support for beans. The beans provide nitrogen in an accessible form that serves as fertilizer for all the crops. The squash, which is low to the ground, inhibits weed growth and maintains soil moisture for the other two partners. Plants growing in polyculture in a Three Sisters garden yield more than if each were grown in monoculture.[5] This Indigenous agricultural practice exemplifies the positive outcomes of reciprocal relationships promoted by diversity. Individuals can perform better in diverse environments than when they attempt to function in isolation or only with others similar to themselves. They—indeed we—are better together. "The lessons of reciprocity are written clearly in a Three Sisters garden," writes Robin Wall Kimmerer, a plant ecologist and enrolled member of the Citizen Potawatomi Nation.[6]

As in all great relationships, timing is crucial for tapping the synergistic potential of the Three Sisters system.[7] Studies of plantain and cassava intercrops corroborate the Three Sisters' lesson that the sequence with which each species is planted and becomes established is crucial for determining the ultimate productivity of polyculture crops.[8]

For the Three Sisters, corn emerges first; the seeds absorb moisture in the soil that promotes germination. The corn seedling takes root, initiates development and expansion of leaves, and establishes robust photosynthesis that allows for a transition into independence. Instead of drawing from food stores in the seed, the seedling will now produce food via photosynthesis. The next sister to emerge is the bean. A sprouting bean that emerges alone grows close to the ground and is highly susceptible to damage and stress from both living and nonliving factors, such as predation or low light. But when coming up next to a corn plant, the bean draws on the support of her corn sister and is elevated—both literally and figuratively. Being lifted from the ground promotes growth. As the stem twines up the corn stalk, the bean has increased exposure to sunlight to drive photosynthesis. As we will see in the next section, the bean's roots also play an important role in providing nitrogen. The third sister, squash, emerges last. The squash plant spreads broad leaves close to the soil surface, seeking the open spaces in the canopy through which light penetrates. More light means more photosynthesis and more production of life-sustaining sugars. The low, sprawling leaves cover and protect the root systems of the first two sisters; they also prevent weeds from becoming established, protect the soil from drying out, and, because they are prickly, deter would-be herbivores of all three sisters.[9] The timing of the sisters' establishment and growth is a well-choreographed dance. The trio embodies, in Kimmerer's words, the "knowledge of relationship," and their dance has implications that spread far beyond their existence and thriving.[10]

When we observe a Three Sisters garden we can easily see how the three plants distribute their leaves in space to avoid competing with each other.[11] But few observers are likely to recognize the underground supporting players in this ecosystem. Roots often form relationships with other organisms in the soil microbiome, and these relationships influence the overall fitness of plants, from establishment to growth to flowering.[12] The Three Sisters system is no exception.

Belowground, the Three Sisters support and complement each other just as well as they do above ground. The roots of corn are rather shallow; they occupy the upper part of the soil, while the deep taproots developed by beans tunnel below them. The squash plant positions roots in places that have not been occupied by the roots of the two previously established sisters. Any place that the stem of a squash plant encounters the soil, the plant can initiate additional roots, known as adventitious roots. These roots, which can be placed in open spaces within the niche, supplement the squash plant's growth and persistence.[13] These adventitious roots, as well as root hairs from the other two sisters, are able to distribute themselves throughout the available parts of the soil, allowing the plants to search for resources and establish relationships with others.[14] This underground cooperation is as important to the sisters' relationship as that above the surface. The reciprocal interactions of the plants in co-cultivation again demonstrate, as Kimmerer puts it, that "all gifts are multiplied in relationship."[15]

In addition to drawing moisture and nutrients from the soil, plant roots establish symbiotic relationships with bacteria and fungi. The bacteria fix nitrogen into a form that plants can use, and fungi form mycorrhizae that improve water uptake and nitrogen and phosphate acquisition. These interactions are not one-way: the plant receives increased access to moisture and fertilizers, while the bacteria and fungi are treated to gifts of sugars from the plant.[16]

In the case of the Three Sisters, the second sister, the bean, provides the nitrogen fertilizer as a result of her colonization by a specific nitrogen-fixing bacterium.[17] Mycorrhizae, although not often a focus of work on the Three Sisters system, also play a critical role, just as they do in natural environments. They are especially important for community building and communication, as a single fungus can connect multiple plants underground, resulting in connections and networks among them. Mycorrhizae not only acquire carbon from the plants they colonize, but they also facilitate the sharing of carbon among the individuals that they interconnect.[18] This kind of interaction results in resource-sharing networks, an economy of sorts, between distinct individuals bound together in community.

Although the Three Sisters work together in harmony, not all interactions in a diverse environment are equally benign. And, so, it's equally important that plants detect and respond accordingly. As we saw in Chapter 2, plants must distinguish whether potential interactions are likely to be beneficial or dangerous—whether others with which they interact are friend or foe. Plants can recognize harmful bacteria—pathogens—by means of specific molecules that are present in the bacterial cell walls. Some of these molecules have been highly conserved through evolution; as a result, many different pathogens contain the same ones. These molecular fragments of pathogens can be detected by plant receptors, thereby serving as a potent signal that danger is imminent.[19] Because these molecules are released when the bacteria interact with the plant surface or with the soil, the signal of a potential invader is sent to neighboring plants, as well. We see this ability to signal danger in some animals, too. When attacked by predators, for example, fish release chemicals that can be smelled by other nearby fish in a group. When those nearby fish are kin, the attacked fish release more of the chemicals.[20]

Plants respond to such threats with defense mechanisms that operate both locally and at a distance. Recall that when attacked by a pathogen, a plant produces volatile organic compounds that travel within the plant itself or through the air, to other plants, warning of danger.

It is this kind of behavior that allows plants to survive and thrive in dynamic conditions. Not only do predators come and go, but soil properties, such as nutrient availability, moisture content, and soil pH, vary, and the composition of plant communities themselves changes with time. Access to light or to nutrients in the soil can change as plants become more densely packed and some grow taller. Heterogeneous environmental conditions can promote ecological community resilience and increase the diversity of ecosystems.[21]

The Three Sisters system shows us that reciprocity in a diverse environment leads to productive growth. It also highlights the beneficial effects of community interactions and suggests the wisdom of an ecosystem-based approach to promoting communication and supporting success. The Three Sisters are also an example of the power of partnering, reciprocal relationships, ecological niche partitioning, and cycling of nutrients or resources.[22] The Sisters' lessons are equally applicable to discussions about communal values.[23]

The greatest and most enduring lesson that may emerge, however, is understanding that each individual in a community brings particular skills and has the potential to offer unique contributions. We must cultivate individual awareness about our distinct contributions, foster synergies among them, and nurture a community that welcomes these gifts and recognizes how they contribute to and elevate the community as a whole.[24]

The Indigenous peoples who developed the Three Sisters crop planting knowledge knew the benefits of planting corn, beans, and squash together long before scientists recognized the reciprocal relationships and named the mechanisms and processes behind them. Think of all the other knowledge that Indigenous groups had and have about the natural world. Perhaps it is time to bridge the gap between Indigenous and scientific bases of knowledge.[25] Bringing together this kind of knowledge in this way mirrors the natural world. The Sisters offer lessons inspired by and transcending plant knowledge. After all, as Kimmerer explains, "science asks us to learn *about* organisms. Traditional knowledge asks us to learn *from* them."[26]

The nature of the reciprocal relationships in the Three Sisters garden can provide guidance for how we humans establish interactions in various domains of life, including personal, professional, and educational. We often view these domains of our existence as competing with each other in terms of time, energy, and resources, among other factors.[27] Because the time and energy we expend in distinct domains is largely driven by perceived rewards and obligations, we tend to view involvement in one domain as taking valuable time and energy away from involvement in the others—leaving us constantly juggling competing demands.

Instead of thinking of these domains as competing with each other, we should consider that integration, or reciprocal cross-feeding of domains, could yield benefits in both personal and professional arenas, just as growing different crops together boosts productivity.[28] As a professor, I often feel torn between my responsibilities to teach, mentor, conduct research, and participate in service activities. As I began to see the overlap in these commitments and to cultivate activities that are synergistic, such as using new discoveries from my research as core materials in my lectures, I gained personal appreciation for the importance of cultivating reciprocity. Indeed, life–career "balance" would result in different priorities and present additional opportunities if we viewed the distinct domains as reciprocal areas of responsibility or opportunity rather than as competitors for time, energy, or resources.

Much like the corn in the Three Sisters garden, the first domain is the foundation that supports the growth of another domain. Having established a strong primary foundation, we may next seek to promote growth of a second domain that is interdependent with and supported by our primary interest.

Finally, we add a third area that is important but lower in priority. Having established the primary criteria by which we wish to evaluate our life or career success, we can assess which complementary activities integrate with or elevate our first and second domains, or "sisters," in ways that result in a partnership. In my professorial career, the three domains are defined by the criteria for review and promotion—research, teaching, and service. In my personal life, major domains such as parenting and work life are defined and the third, such as self-care, is a personal choice. Like the corn, beans, and squash of a Three Sisters garden, these domains are "cooperating, not competing."[29] Taking long walks together with my son in summer is one way that I cooperatively engage the domains of parenting and self-care. The Three Sisters system provides a rich framework for inspiring integration across personal and professional domains.

The Three Sisters also offer us lessons on "the capacity of others [animate beings] as our teachers, as holders of knowledge, as guides," as Kimmerer puts it.[30] These lessons are essential for establishing, promoting, and implementing cross-cultural competence. We are best able to create access for and support the success of individuals from diverse cultural backgrounds when we appreciate the gifts that each of them has to offer. We should put this lesson into practice in many realms—in our communities, our schools, our workplaces.[31] This lesson is becoming increasingly important as the demographics of the US population continue to change and communities of learners and workers rapidly diversify.[32] Our ability to recognize and embrace the reciprocal benefits of diversity is of critical importance.

If we are able to open our eyes to these lessons from the Three Sisters, and indeed from all plants, there is an overflowing abundance of wisdom awaiting our awareness and implementation.

I choose . . . to live so that which came to me as seed goes to the next as blossom, and that which came to me as blossom, goes on as fruit.

—DAWNA MARKOVA,
I Will Not Die an Unlived Life

6

A Plan for Success

I recall my mother watching closely to see when one of her treasured potted plants was reaching the end of its life in a particular container. She would often remark that it would soon be time to repot it or subdivide it. She would carefully dislodge the plant from the old pot and then either settle it into a larger container or separate the offshoots and repot them. Failure to transfer the plant to a location with more abundant resources would cause the plant to atrophy and die or, sometimes, to flower prematurely. As a caregiver, she facilitated the process through careful attention and mediation, helping the plant thrive in its environment while allowing it to naturally move through the next part of its life cycle.

In Chapter 4, we discussed ecological succession in the context of transformation. As we saw, a plant's ability to compete with others or adapt to a changing community determines how long it can survive in a particular environment.[1] If the environment cannot sustain long-term survival, the plant will make a plan for discontinuing its engagement with the current environment. One strategy is to transition from growth to flowering and seed set, in hopes that the seeds will encounter better conditions.

Each plant follows its natural pattern of growth and development based on its history and tuned to its current environment and the community of others with which it coexists. An annual plant must flower and produce seeds in its single season of life or else it misses its opportunity to produce offspring, whereas a perennial plant can afford to miss a season of successful flowering and seed set because it will have opportunities to reproduce in later years.[2] Though plants with different life cycles may exist in the same environment, each has a specific repertoire of behaviors (although environmentally tunable) based on its genetic makeup and must temper its energy expenditure and behavior accordingly.

It is essential for plants to make decisions based on environmental surveillance since, like all organisms, they have a finite amount of energy. They must carefully budget their energy, particularly in times of limited resource availability, since energy used for one activity is no longer available for other activities.

After using their sensory abilities to assess changes in the environment, plants then make decisions about what actions to take to survive and continue to be productive. If deciding that further survival is impossible, the plant is in essence deciding to make plans to support advancement of the next generation.

Plants' responses to their environment are driven by the conditions encountered throughout their life cycle. Early life stages, such as seedling establishment, can influence later life stages, and the way a plant responds to environmental cues at certain life cycle stages impacts its characteristics. Even plants that are genetically very similar can exhibit varying levels of phenotypic plasticity, as mediated by molecular responses to environmental cues. For example, scientists studied two ecotypes—distinct genetic variants—of a small flowering plant, long-stalked starwort (*Stellaria longipes*), that had adapted over many generations to two contrasting environments and responded differently to environmental cues that varied between the divergent habitats.[3] The researchers studied one ecotype that grows in prairies, where there is dense vegetation and shade, and one that grows in alpine meadows, where the vegetation is more sparse and there is less competition for light. The shade-adapted prairie ecotype showed high competitive ability to rapidly elongate in the presence of shade. By contrast, sun-adapted alpine plants showed a much more limited response to shading—they elongated much less when exposed experimentally to light limitation, a signal that they rarely encountered in natural contexts. The observed differential capacities to respond to light availability are driven by interactions between genetic makeup, molecular responses to environmental cues, and environmental history.

The natural habitat, life history, and molecular capacity of a plant to respond to differentially available resources also drive plant responses throughout the life cycle. The effects of environmental history can be observed as early as when the embryonic plant emerges from the seed. This stage of plant growth is known as the seed-to-seedling transition and is a critical stage of plant development that is affected both by the dynamics of the environment in which the plant is rooted and by the environmental history of the population from which it arose.[4] During the seed-to-seedling transition, a critical switch occurs from dependence on stored energy that had been deposited into the embryonic plant by its mother to self-dependent growth fueled by energy produced during photosynthesis. This transition is a precarious one. The seedling must accurately tune its metabolism, making sure to expend its energy carefully so that it accumulates all the components needed to support photosynthesis before its inherited energy stores are depleted. Because seedlings are highly susceptible to predation and other dangers, the seed-to-seedling transition is a bottleneck for species establishment that can determine the composition of plant populations.[5] Although it comprises just a brief portion of a plant's total life cycle, this transition period can drive the dynamics of natural communities and has implications for the maintenance of species diversity. Certainly, general patterns of the timing of germination are determined by evolved life history strategies. However, for many seeds, the timing can be modulated by environmental factors such as light or water availability. Thus, careful regulation of the timing and progression of this transition offers plants a way to manage succession planning in particular environments.[6]

The environment has a profound effect on how plants make the transition from one life stage or generation to the next. Under certain environmental conditions, for example, plants may decide to accelerate their life cycle, or to shed their leaves. Plants do not take a decision to end a life cycle or sacrifice critical plant organs lightly. Yet, they recognize that sacrificing short-term productivity for long-term persistence is sometimes the wisest decision they can make.

Under prolonged shade conditions, some shade-avoiding plants accelerate their development by reducing the amount of time to flowering. A consequence of a shortened life span for an annual plant or shortened season of growth for a perennial is that the amount of time for storing resources is also shortened. Plants that take this path produce fewer and smaller mature seeds.[7] The production of some seeds, however, is presumably better than the risk of continuing in a vegetative, nonreproductive state and not producing any seeds at all if poor conditions persist. In addition to shortening the time to flowering, these plants often reduce their branching, which results in a smaller overall leaf biomass available for energy investment.

Another form of planning for the future is one that we are all familiar with and that brings us great enjoyment: the annual arrival of fall colors. This is a period in which deciduous trees and shrubs drop their leaves to prepare for overwintering. As a central part of this programmed and finely orchestrated process, plants reduce the production of chlorophyll, which is energetically costly, and degrade existing chlorophyll pools. This shuts down the process of photosynthesis, enabling the plant to conserve energy that would be required to maintain photosynthetic apparatuses and avoid the metabolic costs of supporting leaf biomass through the winter. Plants also move nutrients from the leaves to other plant parts that will survive during the cold weather.[8]

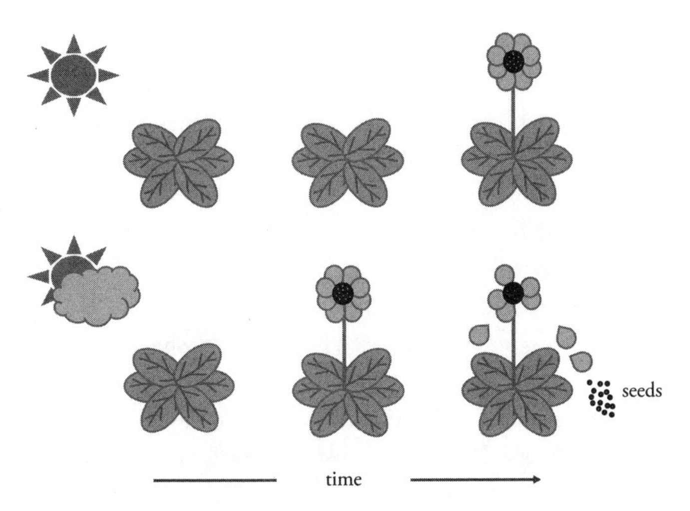

Plants growing in suboptimal conditions, for example in heavy shade (*lower*), can experience limitations to photosynthesis and, thus, energy production, in comparison with those growing in optimal conditions, such as full sun (*upper*). If poor conditions persist, these plants can accelerate flowering to increase the likelihood of producing seeds before the end of their life cycle.

The loss of chlorophyll results in other leaf pigments becoming more visible to the human eye, such as the intense and brilliant yellows and oranges of carotenoids and the reds of anthocyanin.[9] The change in pigment synthesis is coordinated with timing of leaf fall as a function of strategic energy allocation. This process is the basis of a plan for the future in which plants are preparing to exist in a more dormant state. By sacrificing its leaves, a tree can use the minimal energy that is produced or mobilized from carbon stores during the winter months for basal metabolism and processes related to protecting meristems and buds, which will be used to initiate the production of new leaves in the spring. Although leaf drop is distinct in some ways from accelerated flowering, since deciduous trees drop their leaves every year, the timing can be modified to some degree in response to changes in seasonal cues.

Planning for the future can occur at the individual level, as in the examples just described, or it can be coordinated at the community level. One example of community-level planning involves the sharing of resources between mature and younger plants in less than optimal conditions, such as a shortage of resources. Researchers have discovered that in some cases, older plants, called "nurse plants," assist younger, smaller plants (of the same or different species). Although the juvenile plants receive help from the nurse plants, the relationship is reciprocal: both the young and the mature plants grow and survive better together than when growing in isolation, in the same way as the corn, beans, and squash do in the Three Sisters system. The young plants benefit from the shade supplied by the nurse plants, as well as from increased access to water and nutrients provided by the leaf litter deposited beneath the older plants. This leaf litter likely also improves soil properties by modulating soil chemistry and nutrient levels and enhancing symbiotic relationships with bacteria and fungi. These changes in the soil create a feedback loop that supports both the younger and older plants. Another benefit for the nurse plants is that they produce more flowers than do similarly aged plants growing in isolation, possibly because of the improved soil properties. More flowers could attract more pollinators, thus amplifying the effect of a larger floral display on seed set.

Similarly, in forests, older trees can support young trees through active transport of sugars from the mature plant to the younger ones through mycorrhizal networks connecting the plants' roots to meet their robust demands for energy.[10] When the older trees die, they serve as a source of recycled organic components that the younger trees can use for growth and increased fitness.

Additional community responses related to planning for the future and resource sharing are associated with mycorrhizal communities. The fungi that make up mycorrhizae are often associated with a network of different plants.[11] The mycorrhizae allow plants to save energy because they increase nutrient and water uptake by the roots, with the benefit to the plant far outweighing the cost of the sugars shared with the fungal partner.[12] The fact that mycorrhizae connect multiple plants facilitates shared decision-making and community maintenance: plants with excess energy stores can share these with vulnerable community members to support their continued growth and persistence. The extent of this sharing was documented by an ingenious experiment done in a forest in Switzerland. The researchers tracked the carbon (in the form of carbon dioxide) that had been assimilated by a tall spruce tree and found that large amounts of it had been transferred to neighboring trees of different species via a network of mycorrhizae.[13]

Another way that plants support each other and ensure future success is through sending signals to their neighbors when they are attacked. As we saw in Chapter 2, when defending against herbivores, many plants release signals in the form of volatile organic compounds. These signals serve to ward off danger for the plant itself, as well as to warn kin. (Some insects and other herbivores have evolved ways to fight back. These predators release their own signals that disrupt plant-to-plant communication, leaving neighbors confused and in a state that renders them more susceptible to herbivory.[14])

These elaborate responses, which are community based and coordinated at an ecosystem level, generally serve the plants well, at both the individual and the community level. Plants often face multiple stresses at the same time, however, and must prioritize their responses in order to budget their energy appropriately. If a plant is dealing with light stress, for example, it might temporarily suspend its response to other stresses, so that it can prioritize modulating its ability to harvest more light or, in the case of excess light, protect itself from overexcitation.[15] Scientists have also observed that plants dealing with salt stress, such as those growing in saline soils that are increasingly prevalent across the globe, are less able to initiate shade-avoidance responses, and those responding to shade often show reduced abilities to respond to herbivore attack.[16]

In natural communities, plants, as we have seen, have strategies for caring for themselves and engaging with others by budgeting energy, altering their life cycle, sharing resources, or sending danger signals. But when tending our gardens, houseplants, and crops, we humans provide interventions as the caretakers.

We've all had a plant in our care that is just not faring well. So, what can we do about a plant that is experiencing stress? How do we help a houseplant that is not thriving? To problem solve, we generally focus on what is lacking in or wrong with the environment, or, alternatively, what is wrong with the caretaker. We rarely ask whether the plant itself is incapable of growth or demonstrating success.

The caretaker's most common response to a plant that is doing poorly is first and foremost to make a detailed assessment of the plant's environment. Is the plant receiving enough light, or too much light? Does it have the right types and amounts of nutrients? Is the plant being watered too little or too much? Is the temperature too low or too high? Are there signs that pests or herbivores are causing life-threatening damage? Are there other signs of reduced fitness or distress? This kind of thorough analysis of the living and nonliving components of a plant's environment is critical. Typically, a caretaker then considers specific interventions and continues to evaluate the plant's health after these are applied to ensure that the attempts to make things better for the plant are actually working.

When we as caretakers have extensively probed the external environment and have identified specific deficits or unmet requirements, we often recognize that successful plant growth requires new resources or the relocation of existing resources. We assess whether resources already present somewhere in the environment need to be made available to support the plant's growth and development. For example, water might be present in a faucet, but it is of no use if it cannot reach the soil in which the plant is growing. Thorough knowledge of the environment, combined with full awareness of an individual plant's

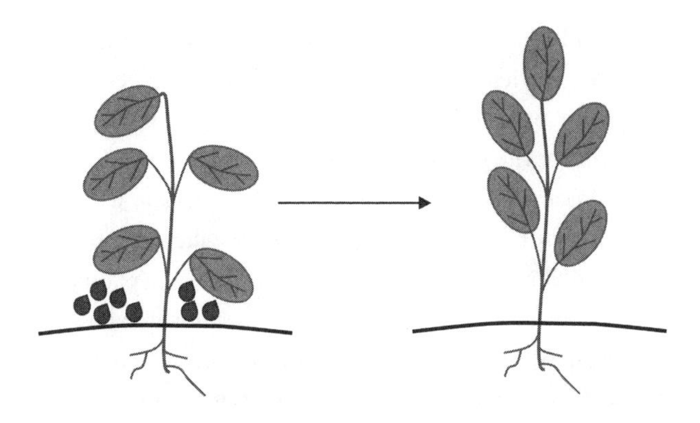

A wilted plant (*left*) clearly needs water, which a caretaker can provide. Such an intervention supports the reestablishment of health for the watered plant (*right*), whereas a wilting plant that does not receive water continues to experience stress and may die.

needs, allow caretakers to connect plants with the particular resources they need to succeed.

In some cases, a resource might be available in sufficient quantities but be deficient in some other way. For example, tap water from the faucet might have impurities that make it substandard. In this case, purification might solve the problem. Alternatively, a different form of water, such as bottled or filtered water, might be needed to support robust plant growth and survival.

To help a plant thrive, the caretaker must be able to recognize its current and evolving needs and then identify and acquire the necessary resources. Given two plants with equal potential for growth, the one with sufficient resources will grow better and exhibit greater productivity than the one with insufficient access to the necessary resources.

As caretakers we frequently seek expert help when our own efforts prove ineffective or when we lack knowledge about what is hindering our plants' growth. We attribute failure to our own caretaking and stewardship inadequacies or inabilities, and as a result we often look for opportunities to improve our care regimen. We may request the assistance of someone we know to be good at taking care of plants. That is, we actively seek training or coaching to learn to become better caretakers, including asking for advice on how to identify what resources the plant needs or how to improve our caretaking abilities.

If a plant is not faring well, the caretaker may, as a very last resort, attribute this outcome to a failure to identify how to facilitate the plant in thriving, but not to a failure on the part of the plant itself. After the caretaker has exhausted all environmental interventions, sought training or called on an expert to intervene, they may ultimately decide that the plant lacks resources that they cannot identify or perhaps is incapable of thriving in a specific environment or under their care. In such instances, however, there is often no negative judgment of the plant per se but rather a reluctant acceptance of the caretaker's personal failure to mediate environmental deficits to support its growth.

To cultivate individual growth and success in people, we must apply the same kind of inquiry-based mindset that we use for plants. I've found that when caretaking a plant, we typically focus on what we can do to help the plant thrive. When mentoring or coaching a person, our inclinations are initially quite different. We are often quick to highlight presumed weaknesses and deficits in the individual rather than seeking to identify environmental factors that might be hindering them.

A comprehensive approach based on a growth mindset is a much more effective way to promote individual development and success. This approach recognizes that it is vital to have a bilateral focus, considering the contributions of both the individual and the environment. Fortunately, in some learning environments, professional work settings, community-based outreach programs, and mentoring programs, mentors and leaders are beginning to explore the impact of environmental factors on individual potential for success or growth, rather than defaulting to deficit-based perspectives.

Despite this progress, much work remains. Just as we do when caring for plants, we should begin our engagement with others by asking systematic questions about the effect of the environment. When plants care for themselves, they sense and perceive signals in the external environment. Their perception then leads to modulation of a plant signaling network, which ultimately leads to an outcome. With other humans, however, we often reverse the process. My work has shown that when we operate from a deficit-based mentality, we move from perceiving a poor outcome to making a negative judgment about the individual. We can be quick to identify weaknesses or assess an individual as incapable of progress when challenges arise. We often default to such judgmental responses rather than asking questions about the individual and their environment.[17]

This tendency is especially pronounced when individuals from minoritized, marginalized, or historically underrepresented and excluded groups encounter challenges acclimating to or succeeding in particular environments.[18] The system often labels them "unable to succeed." Such a deficit-based approach fails to adequately assess the impact of environmental factors on individual outcomes. Those in a position to judge frequently presume that the environment is largely free of detrimental factors that create barriers to success, which may indeed be limiting the potential of the individual. We need to respond the same way we do to plants and expect that individuals are capable of growth. We must then probe many aspects of their surroundings, as well as analyzing how responsive we have been in tending that environment.

We should not, by default, presume that the system in which we and others are trying to advance is infallible or that the environment is adequate. A well-informed understanding of how the system affects an individual will greatly enrich and elevate our engagement practices, from support to mentoring to leadership and advocacy. Additionally, understanding the history of support and inclusion, or lack thereof, in an organization can help decrease susceptibility to the imposter syndrome. This syndrome, which involves feelings of inadequacy despite demonstrated success, is increasingly being recognized as stemming not just from internal factors relating to personality traits, but from external factors such as competition, isolation, and lack of mentorship.[19]

Well-cultivated knowledge about what resources are available, combined with full awareness of individuals' needs, allow us to make targeted efforts to connect individuals with resources to help them succeed. Serving as an environmental steward is one of our most significant roles.[20] And as an effective steward, a growth-minded supporter would recognize when available resources are insufficient and facilitate connection to suitable alternatives or assist in altering an existing resource (in the same way that we would obtain a filter to purify polluted tap water). We can promote this transformation of resources in a community by providing training to improve mentoring, support, and leadership. Community leaders play a key role in setting expectations that community members serve in such supportive roles—providing structures of preparation through support and mentoring, establishing mechanisms of accountability, and rewarding efforts accordingly.

Supporters, mentors, and leaders matter greatly in helping individuals reach their full potential. Given two individuals of equal aptitude, the one connected to the right resources or embedded in an appropriate developmental or support network is much more likely to succeed. The potential for a positive outcome depends, in large part, on caretakers in established networks recognizing how their own experience, expertise, and access to resources can serve the individual needs of those they are supporting and advance their goals. To do so effectively requires ensuring that supporters and colleagues are prepared to offer culturally competent care based on best practices or needed innovations. Community leaders can set explicit goals, as well as provide protected time and incentives to those individuals who seek to elevate their care of others and stewardship of communities.[21]

When supportive relationships are not progressing well, supporters, mentors, and leaders can seek advice from others with experience.[22] An adviser may recommend specific actions to take, assist with increasing awareness of resources, or facilitate connections of those needing support and guidance to available resources.[23] It is not regarded as a weakness to seek advice on plant care from those who are more knowledgeable or successful; in the same way, we need to promote environments, communities, and cultures in which seeking advice about how best to support and mentor others is seen as a strength—indeed, a responsibility—that is encouraged, recognized, and rewarded.

In the unfortunate event that a supportive relationship is not progressing well, the way we care for plants can offer us lessons. Just as we do with a plant that is not thriving, we should consider that the individual offering support may not be meeting the needs of the individual being supported, rather than concluding that the individual has intractable deficits. In the event of mismatches related to culture, in particular, interventions related to "culturally relevant practice" in mentoring and support can improve supporters' abilities to assist people from a wide range of diverse backgrounds.[24] Engaging in such culturally relevant practices can "help us to understand how we can focus on and leverage the wealth of the culture in marginalized and minoritized communities," in addition to assessing systemic structural barriers that impede processes.[25]

One of the primary requirements for a person to function effectively in a culturally relevant support role is to "maintain a dual perspective, seeing [the supported person] as an individual, as well as part of a larger social context."[26] This includes being able to fully comprehend that many of the challenges that individuals from minoritized backgrounds face stem from long-standing histories of systemic inequities.[27] In cases where a supporter is not serving someone well and has already sought advice on how to improve the relationship, it should not be perceived as a failure to admit that the situation is unlikely to end well and to facilitate transfer of the individual to a more suitable caretaker. To do otherwise—to persist in engaging with a person for whom one's specific support skills are not well matched—could lead to complete failure of that person to thrive. The focus should always be on supporting growth—thus, doing no harm—independent of the caretaker's intentions.

We can derive a wealth of knowledge and inspiration about how to support others' growth by observing, contemplating, and enacting lessons based on how we care for plants. The way we engage with plants, taking a growth perspective with an emphasis on inquiry and appreciation, can help transform our engagement with individuals who are being mentored or supported toward a focus on helping them attain their personal and professional goals. Just as we do with plants, we must use cues to guide our care of others to support their success and thriving.

Our personal and professional environments consistently prioritize individualized models of success and achievement over community-based collegiality and reciprocity.[28] We must pivot from deficit-based approaches to mentoring and support, especially of those from groups that are underrepresented, and frequently undersupported, in their learning or professional environment, to growth-based support practices. This growth-based model has great potential to improve outcomes for people from a broad range of demographics and backgrounds. The existence and thriving of individuals in our environment not only enhances their own growth but also has a positive effect on the communities in which they exist, work, and learn.

We have much to learn from how plants care for each other. The way that nurse plants provide benefits to young plants they are "mentoring," and the reciprocal benefits to the nurse in terms of improved growth and reproduction, show us how to prioritize collaboration over competition. These nurse plants, along with the Three Sisters systems, remind us that we thrive better when we work together.

Our capacity and willingness to make good choices and sound decisions is not hardwired in our genes; it is a learned skill, and plants can be great teachers.

—MONICA GAGLIANO, *Thus Spoke the Plant*

CONCLUSION

Groundskeeping

When my son was an infant, we planted "his" tree so that we could watch it grow and mark its seasonal and annual milestones in parallel with his own. The white spruce that we chose is evergreen. While there is no dropping of colorful leaves in the fall, the tree is constantly changing in ways that we reflected on as a family. The first year it was growing a bit off center toward the east, so we carefully tied a string around it, attached the string to a stick on the west side, and gently pulled, to encourage it to grow more upright. Though as a toddler my son didn't fully understand it, we told him that this kind of gentle guidance is often needed in one's youth. And for trees, this action is most effective during the sapling stage, when their trunks are still flexible.

Over the years, as we checked on and attended to his tree, I explained that his exhaled breath, carbon dioxide, would be received as a gift. The tree would take up this essence and convert it to sugars that could help build leaves or even wood, so that he would forever be an integral part of his tree. As the spruce became an "adolescent," we continued to lovingly nurture it and provide any supplemental resources and care it needed. It began to increase its annual growth at the same time as my son, whose ankles suddenly appeared below the hem of pants that had fit perfectly the month before. His tree is still maturing toward adult stage, even as he has made the transition into being a legal adult. He has been its caretaker, and it has been his teacher for nearly two decades; yet we know that there are many more things the tree has to teach him and us.

The previous chapters have introduced you to some of the many lessons that can be learned from plants, which make up a vital if often overlooked component of the natural world. From the enviably pleasant tropical and equatorial climates to seemingly less desirable environments such as desert, alpine, and polar regions, plants can be found growing in nearly any kind of environment. This diversity of habitats is a testament to plants' impressive abilities to perceive what is going on around them, to adapt, and to transform themselves and the environments in which they exist.

Remember, from the very beginning of a plant's existence, a seedling is not just dealing with whatever it finds in a particular space. It must learn to adapt in the niche, or environment, in which it is growing. Ecologically a niche represents the relationship between an organism and its habitat, including the other living things around it. The niche is not fixed, however; through a process called "niche construction," organisms can modify their own and each other's niches by means of their activities and choices.[1] This process of changing an environment primarily for self and to the benefit—or sometimes to the detriment—of others is transformative plant behavior.

Plants continue to learn and adapt throughout their life cycle, constantly weighing energy budget tradeoffs. Defending against damage from a herbivore, for instance, can reduce the energy available for other activities, like expanding leaves or developing new branches. A plant must decide whether to allocate resources to growing or to repelling its foes. A tomato plant whose leaves are being eaten by hornworms uses energy to produce compounds that inhibit the hornworms' growth, but then it has limited energy to contribute to growth and reproduction. Plant responses can also be tuned by cues linked to energy status. If a plant has reduced photosynthetic capacity because of low light conditions, it may be at greater risk of predation because it does not have enough energy to devote to defense.[2] It is from studies of these kinds of situations that we know plants have complex responses to being exposed to multiple cues simultaneously or sequentially.

This kind of sensing, responding, and adapting extends throughout the plant's life cycle, whether it is maximizing potential for capturing light and spinning it into sugars or extending the roots to forage for nutrients. Plants possess powerful abilities to modify their environment to support their own growth as well as that of others—both those with which they share space and time, as well as the generations to follow.

Plants determine when energy is best spent competing, collaborating, or finalizing a life cycle (such as by accelerated flowering when in prolonged shade). They know when an environment does not have the resources required for long-term habitation and successful growth and how to transform it through their own behaviors or through collaboration or reciprocation with others. Pioneer plants, which thrive in disturbed areas, can transform an ecosystem so as to allow for the establishment of other plants. They successfully steward change and cultivate conditions that benefit the next wave of plants.

Lessons emerge not only from plants' behavior in nature, but from our relationship with them. When humans serve as caretakers for plants, we have a growth-based perspective. In caring for our spruce tree, my family and I observe it closely for cues and ask questions about what kind of resources it needs to thrive. We are in tune with its signals so that we can determine if the environment is limiting its flourishing.

Plants provide us with lessons for how to cultivate a robust, sense-driven life. They use sensors to keep close track of what is going on in the environment around them and then use that information to make wise decisions about how to budget their energy, acquire resources, and interact favorably with their neighbors. We can apply these lessons to our own lives, to our mentoring and leadership practices, and to our reciprocal relationships as part of a larger community. Consider that lessons from plants offer us an alternative way of looking at and being in the world, and, for some, a drastically different way to mentor, coach, and lead.

Many people, including mentors and leaders, engage with others as a way to facilitate self-definition, as opposed to functioning from personal agency; they are searching *for* affirmation rather than working *from* affirmation.[3] These individuals are often looking to answer the important questions of who they are, where they should be positioned, and what goals and visions they should be pursuing.

They are attempting to gain a sense of purpose and/or external validation in the course of living, mentoring, or leading, rather than engaging from a place of assurance in which they already have answers to the who, where, and what questions. They need to know who they are and what they offer before they can successfully mentor, coach, and lead others.

When engaging with a community, citizens, mentors, and leaders often display a self-focused, self-affirming perspective rather than aiming to interact reciprocally with others. Self-focused mentoring and leading play out in a form that I call imprinting, that is, training another person to follow one's own behavior or the general norms of a group.[4] In this kind of engagement, mentors and leaders center external validation and promote acculturation; they are seeking confirmation for the choices they have made, including the paths they have traversed and the personal goals they have pursued.[5] This kind of mentoring and leadership is rampant and can have recognizable successes; however, the scope and impact of engaging from a position of seeking purpose (rather than from an already defined purpose) is limited. It is really a personal, internal quest, not a broader vision of purpose.

Alternative viewpoints, processes, and goals are needed. We must envision and work toward an established "vision of purpose" and progress, to what I refer to as an environmentally adaptive mode of living, mentoring, or leading. Like plants, we must learn from our experiences and alter behavior that isn't working. This process begins with critical self-assessment and self-reflection, and a commitment to work from a perspective in which one knows the answers to the who, where, and what questions. Only when you have done so can you live, mentor, and lead effectively. Self-reflection is of key importance for many reasons: it enables you to become aware of your strengths and weaknesses, as well as to clarify your personal goals and aspirations. With this knowledge, you are in a position to identify the right niche for engaging your strengths and finding appropriate opportunities to grow in your areas of weaknesses. Progressive mentors and leaders will promote and model the practice of protecting time for active self-reflection.[6] It is from this place of awareness, a sense of self, that you can move into a position that provides the key niche, or opportunity, to "work out," "grow in," or actualize an individual vision of purpose—to work toward a clearly identified goal.

Plants possess numerous sensors that allow them to monitor what is going on around them and to assess the availability of resources. They can then tune their growth and development to the external environment by means of phenotypic plasticity, which enables them to adjust their responses. They make strategic decisions about how to allocate their resources and can also initiate behaviors that have the potential to transform the environment by altering or increasing the availability of resources.

In human organizations, effective sensors play a key role in detecting areas of needed change, promoting constructive behavior, and facilitating strategic decisions. These individuals are able to quickly detect changes in environmental factors (for example, economic, technological, or competitive factors) or sociocultural factors (for example, social attitudes or political ideology), identify points where interventions could be made, and aid others in implementing these interventions.[7]

Plants assess when they need to be competitive and when it is more prudent to be collaborative. To make this kind of decision, they weigh the energy cost relative to the benefit for improved growth and persistence. For example, although a plant would generally attempt to grow taller than a closely situated neighbor for preferential access to sunlight, if the neighbor is already significantly taller and the race is likely to be lost, the plant will temper its competitive instinct. That is, plants compete only when competition is needed to improve their ability to support their own growth and reproduction and has some likelihood of success. Once competition yields the needed results, they cease competing and shift their energy to living. For plants, competition is about survival, not the thrill of victory.

Humans would benefit from understanding that pursuing competition is a noble cause only when it is required for persistence and thriving. Furthermore, one of the greatest lessons we can learn from plants is that there is power in working together. We must release our overdependence on individualistic success models and instead understand that responses to the environment, whether an office, university, government, or housing community, are usually improved when pursued collectively and collaboratively.

Before investing in a collaborative relationship, plants weigh the costs and benefits. They assess whether sharing the cost of responding to environmental cues and perceiving needs will pay off in terms of increased survival and reproduction. The decision about whether to compete or collaborate is influenced by the presence of kin: a number of studies have shown that collaboration is more likely when neighbors are closely related. Many organisms, including plants, understand that reducing competition or increasing collaboration in the presence of kin has significant implications for survival and thriving of the species as a whole. Humans have a relatively restrictive definition of kin. In addition to our biological kin, we have a tendency to include those whom we treat as having shared value as our functional kin, based on rather narrow definitions of shared ethnicity, race, gender, or socioeconomic status. This perspective affects whom we closely befriend, whom we live together with in neighborhoods and community, and whom we engage with regularly in social contexts. Humans engage with others from similar backgrounds in what I view as a form of kinship, but is a concept more commonly known as homophily.[8] I believe it is time to reconsider our idea of kinship. A key goal for a person in a leadership or mentoring role should be to promote a sense of kinship among all members of a community. Doing so facilitates strategic energy allocation decisions that benefit the whole, rather than particular individuals. Extending this concept out further, we as a human race, and the health and sustaining of our planet, would be well served if we expanded our understanding and enactment of kinship to include all people around the world.

Plants living in communities that contain many different species tend to thrive better and be more productive than those in less diverse communities. Each species occupies a specific niche with a distinct form and mode of being, and together, they can more efficiently make use of light, nutrients, and other resources.

In human environments, we often champion a single path to success in a particular position and hesitate to ask people questions about their own aspirations and vision for personal or professional advancement. Not until we begin to embrace the diversity of the individuals who may come to these positions—the unique experiences, gifts, and skills each person possesses—will we fully experience the wealth of unique "flowers" that each has to offer when encouraged to bloom. While everyone has a role to play in valuing and promoting diverse communities, the responsibility begins with those in positions at the top. To facilitate equity-minded approaches, mentors and leaders must promote cross-cultural competence and culturally aware practices.[9] To do so, they must have a high level of cross-cultural competence themselves.[10] But how do we go about increasing cross-cultural understanding and promoting an inclusive culture of success? Notably, progressive environments that serve a diverse range of individuals usually have community members and leaders who pay close attention to tending the environment: that includes evaluating the climate and identifying barriers, as well as making plans for transformation. The importance of sensing and monitoring the creation and maintenance of supportive, equitable environments is true for community-based, business, and academic organizations alike.[11]

To promote diversity and equity, leaders would do well to remember the lessons of polyculture—the cultivation of diverse plant species together. The Three Sisters system shows us how the community benefits when individuals reciprocally offer their unique capabilities, strengths, and behaviors. Humans are interdependent in ways that we often overlook. If we want more equitable outcomes, we would do well to recognize that everyone benefits when we cultivate people's diverse talents and promote synergies and collaborations among them.

While there are certainly risks in taking alternative paths to success, if we consider the lessons offered to us from plants, we will realize that there may be greater risks in ignoring them. Plants that fail to achieve their ingrained purpose—for example, annual plants that fail to flower in their sole season of existence—risk missing a chance to blossom and leave offspring for future generations. A single individual suffers from this lost opportunity in such an instance, but other inhabitants in the plant's community are also worse off for not having that plant's contribution to community.

Our obsession with championing tried and true paths is likely costing us dearly when we refuse to take the risk of seeing what unique "flowers" each person has to offer. Our communities are enriched by innovation, new modes of thinking, and unique contributions. Yet, welcoming such offerings requires an openness to creativity, inventiveness, and entrepreneurial approaches—especially in professional environments. We must do more than just encourage such trailblazing—we must recognize and reward it.

Trailblazers, whether plants or humans, must be resilient. Plants have the ability to recover from natural disasters such as floods, fires, and hurricanes, as well as manmade disasters, such as the radiation catastrophe at Chernobyl. When we champion resilience, however, we need to ask whether the structures, practices, and indeed the very fabric of our communities demand more resilience and stick-to-itivness from minoritized and marginalized groups than from others.[12] It is essential to consider a person's environmental history and how it may affect their performance and potential for growth and transformation. Our institutions have a history of excluding people from minoritized and marginalized groups and of promoting task-oriented rather than creativity-oriented activities. It takes energy to remain resilient and persevere under these circumstances; thus, mentors and leaders have a responsibility to remove the structural barriers that result in such unequal demands. These barriers differentially affect the ability of individuals to succeed. While resilience is a quality we all should aim for, we must also pay close attention to the equity of the systems in which we are embedded, and look carefully at who is required to be resilient. Leaders who want to cultivate an environment that provides support for a broad range of individuals will be keenly aware of how each person interacts with that environment and will encourage transformative behaviors that promote change where needed.

We would do well in our interactions with others to consider how we take care of plants. For the most part, we begin from the expectation that the plant has the ability to grow and thrive. When the plant is not doing well, we ask questions about the health of the environment (does the plant have enough or too much light?) or about our own abilities as a caretaker (what am I doing wrong?). We do not immediately believe that the plant has deficits.

Unfortunately, when responding to a person who is having difficulty, we often start by asking questions about the individual and why they don't fit well in a particular environment. Such a response is based on the assumption that the fault is with the person rather than the environment, in sharp contrast with what we know to be true from plants. Plants that are otherwise identical can display vastly different outcomes depending on the external environment, such as whether they are grown in the dark or the light. To assess someone's potential for success, we must assess both the negative and the positive influences in their environment. Then we will have a better sense of what adjustments or adaptations are needed to assist those who are struggling.

We can also make good use of the lessons we learn from plants when planning for longer-term environmental changes, such as when considering leadership roles. Perhaps we need pioneers to serve as the first in a progressive line of change agents. These trailblazing individuals provide space and improve access to resources for later waves of leaders with different leadership strengths. Too frequently, we have a one-size-fits-all approach to leadership, rather than understanding that leaders with particular strengths are needed at different times, especially when cultural change is in order. One of the challenges is that we often prioritize long-term presence over long-term outcomes. Pioneer leaders may have a short life in an organization. Yet, if they are successful in opening spaces, establishing new processes, and improving access, it will set the stage for leaders who may be slightly less innovative to enter and have a longer tenure. These long-term leaders who come in second, along with subsequent waves of ecosystem inhabitants, may then proceed to do the valuable work of putting systems into place that provide stable, renewable resources to sustain the community.

This kind of proactive succession planning is important, especially in times of plenty when all seems well. The time to plan is early and often. Plants accomplish this by tracking their individual and community-level success and monitoring energy requirements to achieve important goals, such as reproduction. They follow a plan that allows them to replenish and strategically allocate energy.

Humans also need to function in the present moment and plan ahead for succession. Strategic succession planning requires leaders to lead aptly in the moment, while anticipating future needs and anticipating transitions in leadership. Leaders should act with agility, identifying successors well before they are needed so that they can prepare for the transition. Unfortunately, leaders are often chosen or promoted to maintain the status quo. Until we start promoting sense-driven leadership, we will not see individuals or communities reaching their full potential.

Leaders should play "sensor" roles in their environments, serving as environmental stewards—they should be groundskeepers, not gatekeepers.[13] In this kind of progressive leadership, leaders and mentors show others how to find their niche, how to assess the impact of the environment on growth and behavior, how to address and respond to competition, how to allocate energy to significant endeavors, and how to determine the effect of environmental history on community members. Rather than teaching tactical leadership skills to their successor, the wise leader must cultivate leadership philosophies and vision. This kind of vision is needed to adapt to changing circumstances, and it can also enable leaders to see the potential collaborations and benefits in diverse communities. This approach contrasts with the traditional gatekeeping approach, in which leaders determine who gains access via conceptualizations and assumptions about who can function and thrive in a particular context.[14] Instead, this distinct form of leadership is sense driven and environmentally adaptive; it attends to individuals while at the same time tending the ecosystems in which these individuals exist. I call this form of leadership groundskeeping, in recognition of what we know about the conditions that plants need to successfully thrive.

I have learned so many lessons from plants over the past several decades. I am exceedingly grateful. I also yearn for the time when everyone will live a sense-driven life. Plants show us how to do this. All we have to do is pay attention.

Take a moment and look around you. A plant is surely somewhere within sight. Depending on the time of year or your location on the globe, you may see a seed sprouting, flowers blooming, or the brightly colored leaves of autumn against the sky. All of these behaviors—sprouting, blooming, and changing color—show us how plants are in tune with themselves and their environments, adapting and supporting others from their stationary, yet dynamic, places in the world.

Notes

Acknowledgments

Index

Notes

Introduction

Epigraph: Robin Wall Kimmerer, *Braiding Sweetgrass: Indigenous Wisdom, Scientific Knowledge and the Teachings of Plants* (Minneapolis, MN: Milkweed Editions, 2013), 9.

1. The discussion here focuses on plants that reproduce via seeds. However, some plants, for example, ferns and some mosses, reproduce via spores, whereas others reproduce asexually or clonally through vegetative regeneration from stems, rhizomes (underground stems), bulbs, or tubers; Simon Lei, "Benefits and Costs of Vegetative and Sexual Reproduction in Perennial Plants: A Review of Literature," *Journal of the Arizona-Nevada Academy of Science* 42 (2010): 9–14.

2. James H. Wandersee and Elisabeth E. Schussler, "Preventing Plant Blindness," *American Biology Teacher* 61, no. 2 (1999): 82–86; James H. Wandersee and Elisabeth E. Schussler, "Toward a Theory of Plant Blindness," *Plant Science Bulletin* 17 (2001): 2–9.

3. Sami Schalk, "Metaphorically Speaking: Ableist Metaphors in Feminist Writing," *Disability Studies Quarterly* 33, no. 4 (2013): 3874.

4. Mung Balding and Kathryn J. H. Williams, "Plant Blindness and the Implications for Plant Conservation," *Conservation Biology* 30 (2016): 1192.

5. Balding and Williams, "Plant Blindness"; Caitlin McDonough MacKenzie, Sara Kuebbing, Rebecca S. Barak, et al., "We Do Not Want to 'Cure Plant Blindness' We Want to Grow Plant Love," *Plants, People, Planet* 1, no. 3 (2019): 139–141. Balding and Williams describe "plant blindness" as a "bias" against plants. Their discussion inspired my use of the term "plant bias," as well as my suggestion that decreasing plant bias should lead to increased plant awareness.

6. This bending phenomenon, known as phototropism, was noted in Darwin's treatise on plants: Charles Darwin, *The Power of Movement in Plants* (London: John Murray, 1880), 449. It is controlled by the hormone auxin and has been studied experimentally for a long time, including relatively early work by Briggs and colleagues: Winslow R. Briggs, Richard D. Tocher, and James F. Wilson, "Phototropic Auxin Redistribution in Corn Coleoptiles," *Science* 126, no. 3266 (1957): 210–212.

7. Edward J. Primka and William K. Smith, "Synchrony in Fall Leaf Drop: Chlorophyll Degradation, Color Change, and Abscission Layer Formation in Three Temperate Deciduous Tree Species," *American Journal of Botany* 106, no. 3 (2019): 377–388.

8. Fernando Valladares, Ernesto Gianoli, and José M. Gómez, "Ecological Limits to Plant Phenotypic Plasticity," *New Phytologist* 176 (2007): 749–763.

9. The process by which environmental signals are perceived by sensors within cells and communicated internally is called signal transduction; see Abdul Razaque Memon and Camil Durakovic, "Signal Perception and Transduction in Plants," *Periodicals of Engineering and Natural Sciences* 2, no. 2 (2014): 15–29; Harry B. Smith, "Constructing Signal Transduction Pathways in *Arabidopsis*," *Plant Cell* 11 (1999): 299–301.

10. Sean S. Duffey and Michael J. Stout, "Antinutritive and Toxic Components of Plant Defense against Insects," *Archives of Insect Biochemistry and Physiology* 32 (1996): 3–37.

11. David C. Baulcombe and Caroline Dean, "Epigenetic Regulation in Plant Responses to the Environment," *Cold Spring Harbor Perspectives in Biology* 6 (2014): a019471; Paul F. Gugger, Sorel Fitz-Gibbon, Matteo Pellegrini, and Victoria L. Sork, "Species-wide Patterns of DNA Methylation Variation in *Quercus lobata* and Their Association with Climate Gradients," *Molecular Ecology* 25, no. 8 (2016): 1665–1680; Sonia E. Sultan, "Developmental Plasticity: Re-conceiving the Genotype," *Interface Focus* 7, no. 5 (2017): 20170009.

12. Sun-tracking plants are thought to rotate their leaves and flowers to follow the sun in order to maximize exposure to sunlight or to promote pollinator visits. See M. P. M. Dicker, J. M. Rossiter, I. P. Bond, and P. M. Weaver, "Biomimetic Photo-actuation:

Sensing, Control and Actuation in Sun Tracking Plants," *Bioinspiration & Biomimetics* 9 (2014): 036015; Hagop S. Atamian, Nicky M. Creux, Evan A. Brown, et al., "Circadian Regulation of Sunflower Heliotropism, Floral Orientation, and Pollinator Visits," *Science* 353, no. 6299 (2016): 587–590; Joshua P. Vandenbrink, Evan A. Brown, Stacey L. Harmer, and Benjamin K. Blackman, "Turning Heads: The Biology of Solar Tracking in Sunflower," *Plant Science* 224 (2014): 20–26.

13. Angela Hodge, "Root Decisions," *Plant, Cell & Environment* 32, no. 6 (2009): 628–640; Efrat Dener, Alex Kacelnik, and Hagai Shemesh, "Pea Plants Show Risk Sensitivity," *Current Biology* 26, no. 12 (2016): 1–5.

14. Jason D. Fridley, "Plant Energetics and the Synthesis of Population and Ecosystem Ecology," *Journal of Ecology* 105 (2017): 95–110.

15. Monica Gagliano, Michael Renton, Martial Depczynski, and Stefano Mancuso, "Experience Teaches Plants to Learn Faster and Forget Slower in Environments Where It Matters," *Oecologia* 175, no. 1 (2014): 63–72; Monica Gagliano, Charles I. Abramson, and Martial Depczynski, "Plants Learn and Remember: Lets Get Used to It," *Oecologia* 186, no. 1 (2018): 29–31.

16. Michael Marder, "Plant Intentionality and the Phenomenological Framework of Plant Intelligence," *Plant Signaling & Behavior* 7, no. 11 (2012): 1365–1372.

17. Marder, "Plant Intentionality."

18. For supporters of this view, see Stefano Mancuso and Alessandra Viola, *Brilliant Green: The Surprising History and Science of Plant Intelligence* (Washington,

DC: Island Press, 2015); Paco Calvo, Monica Gagliano, Gustavo M. Souza, and Anthony Trewavas, "Plants Are Intelligent, Here's How," *Annals of Botany* 125, no. 1 (2020): 11–28. For detractors, see Richard Firn, "Plant Intelligence: An Alternative Point of View," *Annals of Botany* 93, no.4 (2004): 345–351; Daniel Kolitz, "Are Plants Conscious?" *Gizmodo,* May 28, 2018, https://gizmodo.com/areplants-conscious-1826365668; Denyse O'Leary, "Scientists: Plants Are NOT Conscious!" *Mind Matters,* July 8, 2019, https://mindmatters.ai/2019/07/scientists-plants-are-not-conscious/. For agnostics, see Daniel A. Chamowitz, "Plants Are Intelligent—Now What," *Nature Plants* 4 (2018): 622–623. For an overview of the debate, see Ephrat Livni, "A Debate over Plant Consciousness Is Forcing Us to Confront the Limitations of the Human Mind," *Quartz,* June 3, 2018, https://qz.com/1294941/a-debate-over-plant-consciousness-isforcing-us-to-confront-the-limitations-of-the-human-mind/.

19. Irwin N. Forseth, and Anne F. Innis, "Kudzu (*Pueraria montana*): History, Physiology, and Ecology Combine to Make a Major Ecosystem Threat," *Critical Reviews in Plant Sciences* 23, no. 5 (2004): 401–413.

1. A Changing Environment

Epigraph: Barbara McClintock, quoted in Evelyn Fox Keller, *A Feeling for the Organism: The Life and Work of Barbara McClintock* (New York: W. H. Freeman, 1983), 199–200.

1. Tomoko Shinomura, "Phytochrome Regulation of Seed Germination," *Journal of Plant Research* 110 (1997): 151–161.

2. Ludwik W. Bielczynski, Gert Schansker, and Roberta Croce, "Effect of Light Acclimation on the Organization of Photosystem II Super- and Sub-Complexes in *Arabidopsis thaliana*," *Frontiers in Plant Science* 7 (2016): 105; N. Friedland, S. Negi, T. Vinogradova-Shah, et al., "Fine-tuning the Photosynthetic Light Harvesting Apparatus for Improved Photosynthetic Efficiency and Biomass Yield," *Scientific Reports* 9 (2019): 13028; Norman P. A. Huner, Gunnar Öquist, and Anastasios Melis, "Photostasis in Plants, Green Algae and Cyanobacteria: The Role of Light Harvesting Antenna Complexes," in *Light-Harvesting Antennas in Photosynthesis*, ed. Beverley Green and William W. Parson (Dordrecht: Springer Netherlands, 2003), 401–421; Beronda L. Montgomery, "Seeing New Light: Recent Insights into the Occurrence and Regulation of Chromatic Acclimation in Cyanobacteria," *Current Opinion in Plant Biology* 37 (2017): 18–23.

3. Tegan Armarego-Marriott, Omar Sandoval Ibañez, and Łucja Kowalewska, "Beyond the Darkness: Recent Lessons from Etiolation and De-etiolation Studies," *Journal of Experimental* Botany 71, no 4 (2020): 1215–1225.

4. Beronda L. Montgomery, "Spatiotemporal Phytochrome Signaling during Photomorphogenesis: From Physiology to Molecular Mechanisms and Back," *Frontiers in Plant Science* 7 (2016): 480; Sookyung Oh, Sankalpi N. Warnasooriya, and Beronda L. Montgomery,

"Downstream Effectors of Light- and Phytochrome-Dependent Regulation of Hypocotyl Elongation in *Arabidopsis thaliana*," *Plant Molecular Biology* 81, no. 6 (2013): 627–640; Sankalpi N. Warnasooriya and Beronda L. Montgomery, "Spatial-Specific Regulation of Root Development by Phytochromes in *Arabidopsis thaliana*," *Plant Signaling & Behavior* 6, no. 12 (2011): 2047–2050.

5. Oh et al., "Downstream Effectors"; Warnasooriya and Montgomery, "Spatial-Specific Regulation."

6. Ariel Novoplansky, "Developmental Plasticity in Plants: Implications of Non-cognitive Behavior," *Evolutionary Ecology* 16, no. 3 (2002): 177–188, 183; Christine M. Palmer, Susan M. Bush, and Julin N. Maloof, "Phenotypic and Developmental Plasticity in Plants," *eLS*, Wiley Online Library, posted June 15, 2012, doi:10.1002/9780470015902.a0002092.pub2.

7. Montgomery, "Spatiotemporal Phytochrome Signaling."

8. Novoplansky, "Developmental Plasticity in Plants"; Stephen C. Stearns, "The Evolutionary Significance of Phenotypic Plasticity: Phenotypic Sources of Variation among Organisms Can Be Described by Developmental Switches and Reaction Norms," *BioScience* 39, no. 7 (1989): 436–445; Palmer et al., "Phenotypic and Developmental Plasticity in Plants."

9. Novoplansky, "Developmental Plasticity in Plants," 179–180.

10. There are, however, limits to the ability to modulate yield and seed set under prolonged stress. M. W.

Adams, "Basis of Yield Component Compensation in Crop Plants with Special Reference to the Field Bean, *Phaseolus vulgaris*," *Crop Science* 7, no. 5 (1967): 505–510.

11. Maaike De Jong and Ottoline Leyser, "Developmental Plasticity in Plants," in *Cold Spring Harbor Symposia on Quantitative Biology*, vol. 77 (Cold Spring Harbor, NY: Cold Spring Harbor Laboratory Press, 2012), 63–73; Stearns, "The Evolutionary Significance of Phenotypic Plasticity."

12. Kerry L. Metlen, Erik T. Aschehoug, and Ragan M. Callaway, "Plant Behavioural Ecology: Dynamic Plasticity in Secondary Metabolites," *Plant, Cell & Environment* 32 (2009): 641–653.

13. Tânia Sousa, Tiago Domingos, J.-C. Poggiale, and S. A. L. M. Kooijman, "Dynamic Energy Budget Theory Restores Coherence in Biology," *Philosophical Transactions of the Royal Society B* 365, no. 1557 (2010): 3413–3428.

14. Fritz Geiser, "Conserving Energy during Hibernation," *Journal of Experimental Biology* 219 (2016): 2086–2087.

15. The ability of plants to change form throughout their life cycle is the observable growth response that is most distinct from mammals, including humans. Ottoline Leyser, "The Control of Shoot Branching: An Example of Plant Information Processing," *Plant, Cell & Environment*, 32, no. 6 (2009): 694–703; Metlen et al., "Plant Behavioural Ecology"; Anthony Trewavas, "What Is Plant Behaviour?" *Plant, Cell & Environment* 32 (2009): 606–616.

16. Carl D. Schlichting, "The Evolution of Phenotypic Plasticity in Plants," *Annual Review of Ecology and Systematics* 17, no. 1 (1986): 667–693; Fernando Valladares, Ernesto Gianoli, and José M. Gómez, "Ecological Limits to Plant Phenotypic Plasticity," *New Phytologist* 176 (2007): 749–763.

17. The movement of petioles to reposition leaves upward is known as hyponasty, whereas downward movement of leaves is called epinasty; these process are regulated by plant hormones such as ethylene and auxin; Jae Young Kim, Young-Joon Park, June-Hee Lee, and Chung-Mo Park, "Developmental Polarity Shapes Thermo-Induced Nastic Movements in Plants," *Plant Signaling & Behavior* 14, no. 8 (2019): 1617609.

18. Sarah Courbier and Ronald Pierik, "Canopy Light Quality Modulates Stress Responses in Plants," *iScience* 22 (2019): 441–452; Diederik H. Keuskamp, Rashmi Sasidharan, and Ronald Pierik, "Physiological Regulation and Functional Significance of Shade Avoidance Responses to Neighbors," *Plant Signaling & Behavior* 5, no. 6 (2010): 655662; Hans de Kroon, Eric J. W. Visser, Heidrun Huber, et al., "A Modular Concept of Plant Foraging Behaviour: The Interplay between Local Responses and Systemic Control," *Plant, Cell & Environment* 32, no. 6 (2009): 704–712.

19. Light-dependent hyponasty, similar to temperature-dependent hyponasty, is driven by changes in cellular turgor pressure or differential growth on one surface of a plant organ, in this case mediated by hormones including ethylene (especially for petioles) and

auxin; Joanna K. Polko, Laurentius A. C. J. Voesenek, Anton J. M. Peeters, and Ronald Pierik, "Petiole Hyponasty: An Ethylene-Driven, Adaptive Response to Changes in the Environment," *AoB Plants* 2011 (2011): plr031.

20. The suppression of lateral branch initiation and growth in the presence of the main or dominant branch is known as apical dominance, which is a hormone-regulated process in plants; Leyser, "The Control of Shoot Branching," 695; Francois F. Barbier, Elizabeth A. Dun, and Christine A. Beveridge, "Apical Dominance," *Current Biology* 27 (2017): R864–R865.

21. David C. Baulcombe and Caroline Dean, "Epigenetic Regulation in Plant Responses to the Environment," *Cold Spring Harbor Perspectives in Biology* 6 (2014): a019471; Sonia E. Sultan, "Developmental Plasticity: Re-Conceiving the Genotype," *Interface Focus* 7, no. 5 (2017): 20170009..

22. Paul F. Gugger, Sorel Fitz-Gibbon, Matteo Pellegrini, and Victoria L. Sork, "Species-Wide Patterns of DNA Methylation Variation in *Quercus lobata* and Their Association with Climate Gradients," *Molecular Ecology* 25, no. 8 (2016): 1665–1680.

23. Quinn M. Sorenson and Ellen I. Damschen, "The Mechanisms Affecting Seedling Establishment in Restored Savanna Understories Are Seasonally Dependent," *Journal of Applied Ecology* 56, no. 5 (2019): 1140–1151.

24. Angela Hodge, "Plastic Plants and Patchy Soils," *Journal of Experimental Botany* 57, no. 2 (2006): 401–411.

25. Angela Hodge, David Robinson, and Alastair Fitter, "Are Microorganisms More Effective than Plants at Competing for Nitrogen?" *Trends in Plant Science* 5, no. 7 (2000): 304–308; Ronald Pierik, Liesje Mommer, and Laurentius A. C. J. Voesenek, "Molecular Mechanisms of Plant Competition: Neighbour Detection and Response Strategies," *Functional Ecology* 27, no. 4 (2013): 841–853.

26. Sultan, "Developmental Plasticity," 3; Brian G. Forde and Pia Walch-Liu, "Nitrate and Glutamate as Environmental Cues for Behavioural Responses in Plant Roots," *Plant, Cell & Environment,* 32, no. 6 (2009): 682–693.

27. Hagai Shemesh, Ran Rosen, Gil Eshel, Ariel Novoplansky, and Ofer Ovadia, "The Effect of Steepness of Temporal Resource Gradients on Spatial Root Allocation," *Plant Signaling & Behavior* 6, no. 9 (2011): 1356–1360.

28. Jocelyn E. Malamy and Katherine S. Ryan, "Environmental Regulation of Lateral Root Initiation in *Arabidopsis,*" *Plant Physiology* 127, no. 3 (2001): 899; Hidehiro Fukaki, and Masao Tasaka, "Hormone Interactions during Lateral Root Formation," *Plant Molecular Biology* 69, no. 4 (2009): 437–449.

29. Xucan Jia, Peng Liu, and Jonathan P. Lynch, "Greater Lateral Root Branching Density in Maize Improves Phosphorus Acquisition for Low Phosphorus Soil," *Journal of Experimental Botany* 69, no. 20 (2018): 4961–4970; Angela Hodge, "Root Decisions," *Plant, Cell & Environment* 32 (2009): 628–640; Angela

Hodge, "The Plastic Plant: Root Responses to Heterogeneous Supplies of Nutrients," *New Phytologist* 162 (2004): 9–24.

30. Xue-Yan Liu, Keisuke Koba, Akiko Makabe, and Cong-Qiang Liu, "Nitrate Dynamics in Natural Plants: Insights Based on the Concentration and Natural Isotope Abundances of Tissue Nitrate," *Frontiers in Plant Science* 5 (2014): 355; Leyser, "The Control of Shoot Branching," 699.

31. Hagai Shemesh, Adi Arbiv, Mordechai Gersani, Ofer Ovadia, and Ariel Novoplansky, "The Effects of Nutrient Dynamics on Root Patch Choice," *PLOS One* 5, no. 5 (2010): e10824; M. Gersani, Z. Abramsky, and O. Falik, "Density-Dependent Habitat Selection in Plants," *Evolutionary Ecology* 12, no. 2 (1998): 223–234; Jia, Liu, and Lynch, "Greater Lateral Root Branching Density in Maize."

32. Beronda L. Montgomery, "Processing and Proceeding," Beronda L. Montgomery website, May 3, 2020, http://www.berondamontgomery.com/writing/processing-and-proceeding/.

2. Friend or Foe

Epigraph: Masaru Emoto, *The Hidden Messages in Water,* trans. David A. Thayne (Hillsboro, OR: Beyond Words Publishing, 2004), 46.

1. Patricia Hornitschek, Séverine Lorrain, Vincent Zoete, et al., "Inhibition of the Shade Avoidance Response by Formation of Non-DNA Binding bHLH

Heterodimers," *EMBO Journal* 28, no. 24 (2009): 3893–3902; Ronald Pierik, Liesje Mommer, and Laurentius A. C. J. Voesenek, "Molecular Mechanisms of Plant Competition: Neighbour Detection and Response Strategies," *Functional Ecology* 27, no. 4 (2013): 841–853; Céline Sorin, Mercè Salla-Martret, Jordi Bou-Torrent, et al., "ATHB4, a Regulator of Shade Avoidance, Modulates Hormone Response in *Arabidopsis* Seedlings," *Plant Journal* 59, no. 2 (2009): 266–277.

2. Adrian G. Dyer, "The Mysterious Cognitive Abilities of Bees: Why Models of Visual Processing Need to Consider Experience and Individual Differences in Animal Performance," *Journal of Experimental Biology* 215, no. 3 (2012): 387–395.

3. Richard Karban and John L. Orrock, "A Judgment and Decision-Making Model for Plant Behavior," *Ecology* 99, no. 9 (2018): 1909–1919; Dimitrios Michmizos and Zoe Hilioti, "A Roadmap towards a Functional Paradigm for Learning and Memory in Plants," *Journal of Plant Physiology* 232 (2019): 209–215.

4. Mieke de Wit, Wouter Kegge, Jochem B. Evers, et al., "Plant Neighbor Detection through Touching Leaf Tips Precedes Phytochrome Signals," *Proceedings of the National Academy of Sciences of the United States of America* 109, no. 36 (2012): 14705–14710.

5. Monica Gagliano, "Seeing Green: The Rediscovery of Plants and Nature's Wisdom," *Societies* 3, no. 1 (2013): 147–157.

6. Richard Karban and Kaori Shiojiri, "Self-Recognition Affects Plant Communication and Defense,"

Ecology Letters 12, no. 6 (2009): 502–506; Richard Karban, Kaori Shiojiri, Satomi Ishizaki, et al., "Kin Recognition Affects Plant Communication and Defence," *Proceedings of the Royal Society B* 280 (2013): 20123062.

7. Amitabha Das, Sook-Hee Lee, Tae Kyung Hyun, et al., "Plant Volatiles as Method of Communication," *Plant Biotechnology Reports* 7, no. 1 (2013): 9–26.

8. Donald F. Cipollini and Jack C. Schultz, "Exploring Cost Constraints on Stem Elongation in Plants Using Phenotypic Manipulation," *American Naturalist* 153, no. 2 (1999): 236–242.

9. Jonathan P. Lynch, "Root Phenes for Enhanced Soil Exploration and Phosphorus Acquisition: Tools for Future Crops," *Plant Physiology* 156, no. 3 (2011): 1041–1049.

10. Ariel Novoplansky, "Picking Battles Wisely: Plant Behaviour under Competition," *Plant, Cell and Environment* 32, no. 6 (2009): 726–741.

11. Michal Gruntman, Dorothee Groß, Maria Májeková, and Katja Tielbörger, "Decision-Making in Plants under Competition," *Nature Communications* 8 (2017): 2235.

12. Changes in energy distribution that occur when a plant is shaded involve a number of hormones, including auxins, which contribute to differential growth, and cytokinins, which arrest leaf development to free up energy resources for growth of stems and petioles. Ethylene and brassinosteroids promote petiole elongation under shade in some plants, whereas abscissic acid inhibits branching. See Diederik H. Keuskamp, Rashmi

Sasidharan, and Ronald Pierik, "Physiological Regulation and Functional Significance of Shade Avoidance Responses to Neighbors," *Plant Signaling & Behavior* 5, no. 6 (2010): 655–662; Pierik et al., "Molecular Mechanisms of Plant Competition"; Chuanwei Yang and Lin Li, "Hormonal Regulation in Shade Avoidance," *Frontiers in Plant Science* 8 (2017): 1527.

13. Irma Roig-Villanova and Jaime Martínez-García, "Plant Responses to Vegetation Proximity: A Whole Life Avoiding Shade," *Frontiers in Plant Science* 7 (2016): 236; Kasper van Gelderen, Chiakai Kang, Richard Paalman, et al., "Far-Red Light Detection in the Shoot Regulates Lateral Root Development through the HY5 Transcription Factor," *Plant Cell* 30, no. 1 (2018): 101–116.

14. Jelmer Weijschedé, Jana Martínková, Hans de Kroon, and Heidrun Huber, "Shade Avoidance in *Trifolium repens:* Costs and Benefits of Plasticity in Petiole Length and Leaf Size," *New Phytologist* 172 (2006): 655–666.

15. M. Franco, "The Influence of Neighbours on the Growth of Modular Organisms with an Example from Trees," *Philosophical Transactions of the Royal Society of London. B, Biological Sciences* 313, no. 1159 (1986): 209–225.

16. Andreas Möglich, Xiaojing Yang, Rebecca A. Ayers, and Keith Moffat, "Structure and Function of Plant Photoreceptors," *Annual Review of Plant Biology* 61 (2010): 21–47; Inyup Paik and Enamul Huq, "Plant Photoreceptors: Multifunctional Sensory Proteins and

Their Signaling Networks," *Seminars in Cell & Developmental Biology* 92 (2019): 114–121.

17. Gruntman et al., "Decision-Making." The plant hormones involved in this process include auxin, gibberellins, and ethylene—the latter well known for its role in the ripening of bananas and apples, described in Lin Ma, and Gang Li, "Auxin-Dependent Cell Elongation during the Shade Avoidance Response," *Frontiers in Plant Science* 10(2019):914 and Ronald Pierik, Eric J.W. Visser, Hans de Kroon, and Laurentius A. C. J. Voesenek, "Ethylene is Required in Tobacco to Successfully Compete with Proximate Neighbours," *Plant, Cell & Environment* 26, no. 8 (2003): 1229–1234.

18. Although there is a general assumption that altruism among kin occurs due to increasing the possibility of passing on one's genes, it is the increased possibility of passing on specific genes, referred to as survival genes or altruism genes, that drives kin selection, rather than bulk gene flow that would include many genes neutral to survival; Justin H. Park, "Persistent Misunderstandings of Inclusive Fitness and Kin Selection: Their Ubiquitous Appearance in Social Psychology Textbooks," *Evolutionary Psychology* 5, no. 4 (2007): 860–873.

19. Guillermo P. Murphy and Susan A. Dudley, "Kin Recognition: Competition and Cooperation in *Impatiens* (Balsaminaceae)," *American Journal of Botany* 96, no. 11 (2009): 1990–1996.

20. María A. Crepy and Jorge J. Casal, "Photoreceptor-Mediated Kin Recognition in Plants," *New Phytologist* 205, no. 1 (2015): 329–338; Murphy and Dudley, "Kin Recognition."

21. Heather Fish, Victor J. Lieffers, Uldis Silins, and Ronald J. Hall, "Crown Shyness in Lodgepole Pine Stands of Varying Stand Height, Density, and Site Index in the Upper Foothills of Alberta," *Canadian Journal of Forest Research* 36, no. 9 (2006): 2104–2111; Francis E. Putz, Geoffrey G. Parker, and Ruth M. Archibald, "Mechanical Abrasion and Intercrown Spacing," *American Midland Naturalist* 112, no. 1 (1984): 24–28

22. Franco, "The Influence of Neighbours on the Growth of Modular Organisms"; Alan J. Rebertus, "Crown Shyness in a Tropical Cloud Forest," *Biotropica* vol. 20, no. 4 (1988): 338–339.

23. Tomáš Herben and Ariel Novoplansky, "Fight or Flight: Plastic Behavior under Self-Generated Heterogeneity," *Evolutionary Ecology* 24, no. 6 (2010): 1521–1536.

24. Mieke de Wit, Gavin M. George, Yetkin Çaka Ince, et al., "Changes in Resource Partitioning Between and Within Organs Support Growth Adjustment to Neighbor Proximity in *Brassicaceae* Seedlings," *Proceedings of the National Academy of Sciences of the United States of America* 115, no. 42 (2018): E9953–E9961; Charlotte M. M. Gommers, Sara Buti, Danuše Tarkowská, et al., "Organ-Specific Phytohormone Synthesis in Two *Geranium* Species with Antithetical Responses to

Far-red Light Enrichment," *Plant Direct* 2 (2018): 1–12; Yang and Li, "Hormonal Regulation in Shade Avoidance."

25. S. Mathur, L. Jain, and A. Jajoo, "Photosynthetic Efficiency in Sun and Shade Plants," *Photosynthetica* 56, no. 1 (2018): 354–365.

26. Crepy and Casal, "Photoreceptor-Mediated Kin Recognition"; Gruntman et al., "Decision-making."

27. Robert Axelrod and William D. Hamilton, "The Evolution of Cooperation," *Science* 211, no. 4489 (1981): 1390–1396.

28. Joseph M. Craine and Ray Dybzinski, "Mechanisms of Plant Competition for Nutrients, Water and Light," *Functional Ecology* 27, no. 4 (2013): 833–840; M. Gersani, Z. Abramsky, and O. Falik, "Density-Dependent Habitat Selection in Plants," *Evolutionary Ecology* 12, no. 2 (1998): 223–234.

29. H. Marschner and V. Römheld, "Strategies of Plants for Acquisition of Iron," *Plant and Soil* 165, no. 2 (1994): 261–274; Ricardo F. H. Giehl and Nicolaus von Wirén, "Root Nutrient Foraging," *Plant Physiology* 166, no. 2 (2014): 509–517; Daniel P. Schachtman, Robert J. Reid, and Sarah M. Ayling, "Phosphorus Uptake by Plants: From Soil to Cell," *Plant Physiology* 116, no. 2 (1998): 447–453.

30. Felix D. Dakora and Donald A. Phillips, "Root Exudates as Mediators of Mineral Acquisition in Low-nutrient Environments," *Plant and Soil* 245 (2002): 35–47; Jordan Vacheron, Guilhem Desbrosses, Marie-Lara Bouffaud, et al., "Plant Growth-promoting Rhizo-

bacteria and Root System Functioning," *Frontiers in Plant Science* 4 (2013): 356.

31. H. Jochen Schenk, "Root Competition: Beyond Resource Depletion," *Journal of Ecology* 94, no. 4 (2006): 725–739.

32. Susan A. Dudley and Amanda L. File, "Kin Recognition in an Annual Plant," *Biology Letters* 3, no. 4 (2007): 435–438. Such responses are often associated with competition being affected by the "input-matching rule," which states that the amount of available resources, or energy input, influences behavior that can be adjusted depending on the presence of kin or non-kin competitors; see Geoffrey A. Parker, "Searching for Mates," in *Behavioural Ecology: An Evolutionary Approach,* ed. John R. Krebs and Nicholas B. Davies (Oxford: Blackwell Scientific, 1978), 214–244.

33. Meredith L. Biedrzycki, Tafari A. Jilany, Susan A. Dudley, and Harsh P. Bais, "Root Exudates Mediate Kin Recognition in Plants," *Communicative and Integrative Biology* 3, no. 1 (2010): 28–35.

34. Richard Karban, Louie H. Yang, and Kyle F. Edwards, "Volatile Communication between Plants That Affects Herbivory: A Meta-Analysis," *Ecology Letters* 17, no. 1 (2014): 44–52.

35. Justin B. Runyon, Mark C. Mescher, and Consuelo M. De Moraes, "Volatile Chemical Cues Guide Host Location and Host Selection by Parasitic Plants," *Science* 313, no. 5795 (2006): 1964–1967.

36. Kathleen L Farquharson, "A Sesquiterpene Distress Signal Transmitted by Maize," *Plant Cell* 20, no. 2

(2008): 244; Pierik et al., "Molecular Mechanisms of Plant Competition," 844.

37. Robin Wall Kimmerer, *Braiding Sweetgrass: Indigenous Wisdom, Scientific Knowledge and the Teachings of Plants* (Minneapolis, MN: Milkweed Editions, 2015), 133; Janet I. Sprent, "Global Distribution of Legumes," in *Legume Nodulation: A Global Perspective* (Oxford: Wiley-Blackwell, 2009), 35–50; Jungwook Yang, Joseph W. Kloepper, and Choong-Min Ryu, "Rhizosphere Bacteria Help Plants Tolerate Abiotic Stress," *Trends in Plant Science* 14, no. 1 (2009): 1–4; Sally E. Smith and David Read, "Introduction," in *Mycorrhizal Symbiosis,* 3rd ed. (London: Academic Press, 2008), 1–9.

38. Yina Jiang, Wanxiao Wang, Qiujin Xie, et al., "Plants Transfer Lipids to Sustain Colonization by Mutualistic Mycorrhizal and Parasitic Fungi," *Science* 356, no. 6343 (2017): 1172–1175; Andreas Keymer, Priya Pimprikar, Vera Wewer, et al., "Lipid Transfer From Plants to Arbuscular Mycorrhiza Fungi," *eLIFE* 6 (2017): e29107; Leonie H. Luginbuehl, Guillaume N. Menard, Smita Kurup, et al., "Fatty Acids in Arbuscular Mycorrhizal Fungi Are Synthesized by the Host Plant," *Science* 356, no. 6343 (2017): 1175–1178; Tamir Klein, Rolf T. W. Siegwolf, and Christian Körner, "Belowground Carbon Trade among Tall Trees in a Temperate Forest," *Science* 352, no. 6283 (2016): 342–344.

39. Mathilde Malbreil, Emilie Tisserant, Francis Martin, and Christophe Roux, "Genomics of Arbuscular Mycorrhizal Fungi: Out of the Shadows," *Advances in Botanical Research* 70 (2014): 259–290.

40. Zdenka Babikova, Lucy Gilbert, Toby J. A. Bruce, et al., "Underground Signals Carried through Common Mycelial Networks Warn Neighbouring Plants of Aphid Attack," *Ecology Letters* 16, no. 7 (2013): 835–843.

41. Amanda L. File, John Klironomos, Hafiz Maherali, and Susan A. Dudley, "Plant Kin Recognition Enhances Abundance of Symbiotic Microbial Partner," *PLOS One* 7, no. 9 (2012): e45648.

42. Angela Hodge, "Root Decisions," *Plant, Cell & Environment* 32 (2009): 628–640. .

43. Tereza Konvalinková and Jan Jansa, "Lights Off for Arbuscular Mycorrhiza: On Its Symbiotic Functioning under Light Deprivation," *Frontiers in Plant Science* 7 (2016): 782.

44. Abeer Hashem, Elsayed F. Abd_Allah, Abdulaziz A. Alqarawi, et al., "The Interaction between Arbuscular Mycorrhizal Fungi and Endophytic Bacteria Enhances Plant Growth of *Acacia gerrardii* under Salt Stress," *Frontiers in Microbiology* 7 (2016): 1089.

45. Pedro M. Antunes, Amarilis De Varennes, Istvan Rajcan, and Michael J. Goss, "Accumulation of Specific Flavonoids in Soybean (*Glycine max* (L.) Merr.) as a Function of the Early Tripartite Symbiosis with Arbuscular Mycorrhizal Fungi and *Bradyrhizobium japonicum* (Kirchner) Jordan," *Soil Biology and Biochemistry* 38, no. 6 (2006): 1234–1242; Sajid Mahmood Nadeem, Maqshoof Ahmad, Zahir Ahmad Zahir, et al., "The Role of Mycorrhizae and Plant Growth Promoting Rhizobacteria (PGPR) in Improving Crop Productivity under

Stressful Environments," *Biotechnology Advances* 32, no. 2 (2014): 429–448.

46. Individual success models are described in Joseph A. Whittaker and Beronda L. Montgomery, "Cultivating Diversity and Competency in STEM: Challenges and Remedies for Removing Virtual Barriers to Constructing Diverse Higher Education Communities of Success," *Journal of Undergraduate Neuroscience Education* 11, no. 1 (2012): A44–A51; Beronda L. Montgomery, Jualynne E. Dodson, and Sonya M. Johnson, "Guiding the Way: Mentoring Graduate Students and Junior Faculty for Sustainable Academic Careers," *SAGE Open* 4, no. 4 (2014): doi: 10.1177/2158244014558043.

47. Patricia Matthew, ed., *Written/Unwritten: Diversity and the Hidden Truths of Tenure*. (Chapel Hill: University of North Carolina Press, 2016).

3. Risk to Win

Epigraph: Hope Jahren, *Lab Girl* (New York: Knopf, 2016), 52.

1. Janice Friedman and Matthew J. Rubin, "All in Good Time: Understanding Annual and Perennial Strategies in Plants," *American Journal of Botany* 102, no. 4 (2015): 497–499.

2. Corrine Duncan, Nick L. Schultz, Megan K. Good, et al., "The Risk-Takers and -Avoiders: Germination Sensitivity to Water Stress in an Arid Zone with Unpredictable Rainfall," *AoB Plants* 11, no. (2019): plz066.

3. Thomas Caraco, Steven Martindale, and Thomas S. Whittam, "An Empirical Demonstration of Risk-Sensitive Foraging Preferences," *Animal Behaviour* 28, no. 3 (1980): 820–830; Hiromu Ito, "Risk Sensitivity of a Forager with Limited Energy Reserves in Stochastic Environments," *Ecological Research* 34, no. 1 (2019): 9–17; Alex Kacelnik, and Melissa Bateson, "Risk-sensitivity: Crossroads for Theories of Decision-making," *Trends in Cognitive Sciences* 1, no. 8 (1997): 304–309.

4. Richard Karban, John L. Orrock, Evan L. Preisser, and Andrew Sih, "A Comparison of Plants and Animals in Their Responses to Risk of Consumption," *Current Opinion in Plant Biology* 32 (2016): 1–8.

5. Efrat Dener, Alex Kacelnik, and Hagai Shemesh, "Pea Plants Show Risk Sensitivity," *Current Biology* 26, no. 13 (2016): 1763–1767; Hagai Shemesh, Adi Arbiv, Mordechai Gersani, et al., "The Effects of Nutrient Dynamics on Root Patch Choice," *PLOS One* 5, no. 5 (2010): e10824.

6. Hagai Shemesh, Ran Rosen, Gil Eshel, et al., "The Effect of Steepness of Temporal Resource Gradients on Spatial Root Allocation," *Plant Signaling & Behavior* 6, no. 9 (2011): 1356–1360.

7. Shemesh et al., "The Effects of Nutrient Dynamics"; Hagai Shemesh and Ariel Novoplansky, "Branching the Risks: Architectural Plasticity and Bet-hedging in Mediterranean Annuals," *Plant Biology* 15, no. 6 (2013): 1001–1012.

8. Enrico Pezzola, Stefano Mancuso, and Richard Karban, "Precipitation Affects Plant Communication and Defense," *Ecology* 98, no. 6 (2017): 1693–1699.

9. Omer Falik, Yonat Mordoch, Lydia Quansah, et al., "Rumor Has It . . . : Relay Communication of Stress Cues in Plants," *PLOS One* 6, no. 11 (2011): e23625.

10. Chuanwei Yang and Lin Li, "Hormonal Regulation in Shade Avoidance," *Frontiers in Plant Science* 8 (2017): 1527.

11. Virginia Morell, "Plants Can Gamble," *Science Magazine News,* June 2016, http://www.sciencemag.org/news/2016/06/plants-can-gamble-according-study.

12. Dener, Kacelnik, and Shemesh, "Pea Plants Show Risk Sensitivity."

13. Stefan Hörtensteiner and Bernhard Kräutler, "Chlorophyll Breakdown in Higher Plants," *Biochimica et Biophysica Acta (BBA)-Bioenergetics* 1807, no. 8 (2011): 977–988; Hazem M. Kalaji, Wojciech Bąba, Krzysztof Gediga, et al., "Chlorophyll Fluorescence as a Tool for Nutrient Status Identification in Rapeseed Plants," *Photosynthesis Research* 136, no. 3 (2018): 329–343; Angela Hodge, "Root Decisions," *Plant, Cell & Environment* 32, no. 6 (2009): 630.

14. Hodge, "Root Decisions," 629.

15. Bagmi Pattanaik, Andrea W. U. Busch, Pingsha Hu, Jin Chen, and Beronda L. Montgomery, "Responses to Iron Limitation Are Impacted by Light Quality and Regulated by RcaE in the Chromatically Acclimating Cyanobacterium *Fremyella diplosiphon,*" *Microbiology* 160, no. 5 (2014): 992–1005; Sigal Shcolnick and Nir Keren, "Metal Homeostasis in Cyanobacteria and Chloroplasts. Balancing Benefits and Risks to

the Photosynthetic Apparatus," *Plant Physiology* 141, no. 3 (2006): 805–810.

16. W. L. Lindsay and A. P. Schwab, "The Chemistry of Iron in Soils and Its Availability to Plants," *Journal of Plant Nutrition* 5, no. 4–7 (1982): 821–840.

17. Tristan Lurthy, Cécile Cantat, Christian Jeudy, et al., "Impact of Bacterial Siderophores on Iron Status and Ionome in Pea," *Frontiers in Plant Science* 11 (2020): 730.

18. H. Marschner and V. Römheld, "Strategies of Plants for Acquisition of Iron," *Plant and Soil* 165, no. 2 (1994): 261–274.

19. Lurthy et al., "Impact of Bacterial Siderophores."

20. Chong Wei Jin, Yi Quan Ye, and Shao Jian Zheng, "An Underground Tale: Contribution of Microbial Activity to Plant Iron Acquisition via Ecological Processes," *Annals of Botany* 113, no. 1 (2014): 7–18.

21. Shah Jahan Leghari, Niaz Ahmed Wahocho, Ghulam Mustafa Laghari, et al., "Role of Nitrogen for Plant Growth and Development: A Review," *Advances in Environmental Biology* 10, no. 9 (2016): 209–219.

22. Philippe Nacry, Eléonore Bouguyon, and Alain Gojon, "Nitrogen Acquisition by Roots: Physiological and Developmental Mechanisms Ensuring Plant Adaptation to a Fluctuating Resource," *Plant and Soil* 370, no. 1–2 (2013): 1–29.

23. Ricardo F. H. Giehl and Nicolaus von Wirén, "Root Nutrient Foraging," *Plant Physiology* 166, no. 2 (2014): 509–517.

24. Nitrogen-fixing bacteria such as *Rhizobia* and *Frankia* are housed in nodules inside plant roots (most commonly those of leguminous plants such as beans), while other nitrogen-fixing organisms, such as cyanobacteria, can be housed either on the external surface of roots or internally. For reviews, see Claudine Franche, Kristina Lindström, and Claudine Elmerich, "Nitrogen-Fixing Bacteria Associated with Leguminous and Non-Leguminous Plants," *Plant and Soil* 321, no. 1–2 (2009): 35–59; Florence Mus, Matthew B. Crook, Kevin Garcia, et al., "Symbiotic Nitrogen Fixation and the Challenges to Its Extension to Nonlegumes," *Applied and Environmental Microbiology* 82, no. 13 (2016): 3698–3710; Carole Santi, Didier Bogusz, and Claudine Franche, "Biological Nitrogen Fixation in Non-Legume Plants," *Annals of Botany* 111, no. 5 (2013): 743–767.

25. Philippe Hinsinger, "Bioavailability of Soil Inorganic P in the Rhizosphere as Affected by Root-Induced Chemical Changes: A Review," *Plant and Soil* 237 (2001): 173–195.

26. Daniel P. Schachtman, Robert J. Reid, and Sarah M. Ayling, "Phosphorus Uptake by Plants: From Soil to Cell," *Plant Physiology* 116, no. 2 (1998): 447–453.

27. Alan E. Richardson, Jonathan P. Lynch, Peter R. Ryan, et al., "Plant and Microbial Strategies to Improve the Phosphorus Efficiency of Agriculture," *Plant and Soil* 349 (2011): 121–156; Schachtman et al., "Phosphorus Uptake by Plants."

28. Carroll P. Vance, Claudia Uhde-Stone, and Deborah L. Allan, "Phosphorus Acquisition and Use: Critical Adaptations by Plants for Securing a Nonrenewable Resource," *New Phytologist* 157, no. 3 (2003): 423–447.

29. K. G. Raghothama, "Phosphate Acquisition," *Annual Review of Plant Biology* 50, no. 1 (1999): 665–693; Schachtman et al., "Phosphorus Uptake by Plants"; Marcel Bucher, "Functional Biology of Plant Phosphate Uptake at Root and Mycorrhiza Interfaces," *New Phytologist* 173, no. 1 (2007): 11–26.

30. Martina Friede, Stephan Unger, Christine Hellmann, and Wolfram Beyschlag, "Conditions Promoting Mycorrhizal Parasitism Are of Minor Importance for Competitive Interactions in Two Differentially Mycotrophic Species," *Frontiers in Plant Science* 7 (2016): 1465.

31. Eiji Gotoh, Noriyuki Suetsugu, Takeshi Higa, et al., "Palisade Cell Shape Affects the Light-Induced Chloroplast Movements and Leaf Photosynthesis," *Scientific Reports* 8, no. 1 (2018): 1–9; L. A. Ivanova and V. I. P'yankov, "Structural Adaptation of the Leaf Mesophyll to Shading," *Russian Journal of Plant Physiology* 49, no. 3 (2002): 419–431.

32. Photoprotective pigments, including xanthophylls and anthocyanins, are more abundant in sun leaves than in shade leaves. Investing in such proteins is energetically costly. See J. A. Gamon and J. S. Surfus, "Assessing Leaf Pigment Content and Activity with a Reflectometer," *New Phytologist* 143, no. 1 (1999): 105–117;

Susan S. Thayer and Olle Björkman, "Leaf Xanthophyll Content and Composition in Sun and Shade Determined by HPLC," *Photosynthesis Research* 23, no. 3 (1990): 331–343.

33. Shemesh and Novoplansky, "Branching the Risks"; Hagai Shemesh, Benjamin Zaitchik, Tania Acuña, and Ariel Novoplansky, "Architectural Plasticity in a Mediterranean Winter Annual," *Plant Signaling & Behavior* 7, no. 4 (2012): 492–501.

34. Nir Sade, Alem Gebremedhin, and Menachem Moshelion, "Risk-taking Plants: Anisohydric Behavior as a Stress-resistance Trait," *Plant Signaling & Behavior* 7, no.7 (2012): 767–770.

4. Transformation

Epilogue: Amy Leach, *Things That Are* (Minneapolis, MN: Milkweed Editions, 2012), 40.

1. Eric Wagner, *After the Blast: The Ecological Recovery of Mount St. Helens* (Seattle: University of Washington Press, 2020).

2. Garrett A. Smathers and Dieter Mueller-Dombois, *Invasion and Recovery of Vegetation after a Volcanic Eruption in Hawaii* (Washington, DC: National Park Service, 1974); Gregory H. Aplet, R. Flint Hughes, and Peter M. Vitousek, "Ecosystem Development on Hawaiian Lava Flows: Biomass and Species Composition," *Journal of Vegetation Science* 9, no. 1 (1998): 17–26.

3. Leigh B. Lentile, Penelope Morgan, Andrew T. Hudak, et al., “Post-fire Burn Severity and Vegetation Response Following Eight Large Wildfires across the Western United States,” *Fire Ecology* 3, no. 1 (2007): 91–108.

4. Lentile et al., “Post-fire Burn Severity”; Diane H. Rachels, Douglas A. Stow, John F. O’Leary, et al., “Chaparral Recovery Following a Major Fire with Variable Burn Conditions,” *International Journal of Remote Sensing* 37, no. 16 (2016): 38363857.

5. For examples see A. J. Kayll and C. H. Gimingham, “Vegetative Regeneration of *Calluna vulgaris* after Fire,” *Journal of Ecology* 53, no. 3 (1965): 729–734; Nandita Mondal and Raman Sukumar, “Regeneration of Juvenile Woody Plants after Fire in a Seasonally Dry Tropical Forest of Southern India,” *Biotropica* 47, no. 3 (2015): 330–338; Stephen J. Pyne, “How Plants Use Fire (and Are Used by It),” *Fire Wars,* Nova online, PBS, June 2002, https://www.pbs.org/wgbh/nova/fire/plants.html.

6. Timothy A. Mousseau, Shane M. Welch, Igor Chizhevsky, et al., “Tree Rings Reveal Extent of Exposure to Ionizing Radiation in Scots Pine *Pinus sylvestris,*” *Trees* 27, no. 5 (2013): 1443–1453.

7. Nicholas A. Beresford, E. Marian Scott, and David Copplestone, “Field Effects Studies in the Chernobyl Exclusion Zone: Lessons to Be Learnt,” *Journal of Environmental Radioactivity* 211 (2020): 105893.

8. Gordon C. Jacoby and Rosanne D. D’Arrigo, “Tree Rings, Carbon Dioxide, and Climatic Change,”

Proceedings of the National Academy of Sciences 94, no. 16 (1997): 8350–8353.

9. Christophe Plomion, Grégoire Leprovost, and Alexia Stokes, "Wood Formation in Trees," *Plant Physiology* 127, no. 4 (2001): 1513–1523; Keith Roberts and Maureen C. McCann, "Xylogenesis: The Birth of a Corpse," *Current Opinion in Plant Biology* 3, no. 6 (2000): 517–522.

10. Veronica De Micco, Marco Carrer, Cyrille B. K. Rathgeber, et al., "From Xylogenesis to Tree Rings: Wood Traits to Investigate Tree Response to Environmental Changes," *IAWA Journal* 40, no. 2 (2019): 155–182; Jacoby and D'Arrigo, "Tree Rings."

11. Mousseau et al., "Tree Rings Reveal Extent of Exposure," 1443.

12. Timothy A. Mousseau, Gennadi Milinevsky, Jane Kenney-Hunt, and Anders Pape Møller, "Highly Reduced Mass Loss Rates and Increased Litter Layer in Radioactively Contaminated Areas," *Oecologia* 175, no. 1 (2014): 429–437.

13. Igor Kovalchuk, Vladimir Abramov, Igor Pogribny, and Olga Kovalchuk, "Molecular Aspects of Plant Adaptation to Life in the Chernobyl Zone," *Plant Physiology* 135, no. 1 (2004): 357–363.

14. Cynthia C. Chang and Benjamin L. Turner, "Ecological Succession in a Changing World," *Journal of Ecology* 107, no. 2 (2019): 503–509; Karel Prach and Lawrence R. Walker, "Differences between Primary and Secondary Plant Succession among Biomes of the World," *Journal of Ecology* 107, no. 2 (2019): 510–516. The lesser

degree of severity during secondary succession refers to a lesser impact on the environment compared to primary succession, rather than the impact on individuals. Devastating forest fires can result in complete displacement and homelessness for animals and humans, which is certainly felt as a severe disturbance to those involved.

15. Chang and Turner, "Ecological Succession in a Changing World."

16. Karel Prach and Lawrence R. Walker, "Four Opportunities for Studies of Ecological Succession," *Trends in Ecology & Evolution* 26, no. 3 (2011): 119–123.

17. Prach and Walker, "Four Opportunities for Studies of Ecological Succession," 120.

18. Malcolm J. Zwolinski, "Fire Effects on Vegetation and Succession," in *Proceedings of the Symposium on Effects of Fire Management on Southwestern Natural Resources* (Fort Collins, CO: USDA-Forest Service, 1990), 18–24. Here, colonization refers to the biological process of plants establishing themselves in an ecological niche. In drawing lessons from plants in this context, direct correlation to human colonization, which is often associated with appropriation of both land and culture, is not intended in any way.

19. I. R. Noble and R. O. Slatyer, "The Use of Vital Attributes to Predict Successional Changes in Plant Communities Subject to Recurrent Disturbances," *Vegetatio* 43, no. 1/2 (1980): 5–21; Zwolinski, "Fire Effects on Vegetation and Succession," 22.

20. Joseph H. Connell and Ralph O. Slatyer, "Mechanisms of Succession in Natural Communities and Their

Role in Community Stability and Organization," *American Naturalist* 111, no. 982 (1977): 1119–1144.

21. Connell and Slatyer, "Mechanisms of Succession"; Tiffany M. Knight and Jonathan M. Chase, "Ecological Succession: Out of the Ash," *Current Biology* 15, no. 22 (2005): R926–R927.

22. Knight and Chase, "Ecological Succession," R926.

23. Mark E. Ritchie, David Tilman, and Johannes M. H. Knops, "Herbivore Effects on Plant and Nitrogen Dynamics in Oak Savanna," *Ecology* 79, no. 1 (1998): 165–177.

24. Peter M. Vitousek, Pamela A. Matson, and Keith Van Cleve, "Nitrogen Availability and Nitrification during Succession: Primary, Secondary, and Old-Field Seres," *Plant Soil* 115 (1989): 233; Jonathan J. Halvorson, Eldon H. Franz, Jeffrey L. Smith, and R. Alan Black, "Nitrogenase Activity, Nitrogen Fixation, and Nitrogen Inputs by Lupines at Mount St. Helens," *Ecology* 73, no. 1 (1992): 87–98; Henrik Hartmann and Susan Trumbore, "Understanding the Roles of Nonstructural Carbohydrates in Forest Trees—From What We Can Measure to What We Want to Know," *New Phytologist* 211, no. 2 (2016): 386–403; Robin Wall Kimmerer, *Braiding Sweetgrass: Indigenous Wisdom, Scientific Knowledge and the Teachings of Plants* (Minneapolis, MN: Milkweed Editions, 2015), 133; Knight and Chase, "Ecological Succession," R926; Janet I. Sprent, "Global Distributions of Legumes," in *Legume Nodulation: A Global Perspective* (Oxford: Wiley-Blackwell, 2009),

35–50; Jungwook Yang, Joseph W. Kloepper, and Choong-Min Ryu, "Rhizosphere Bacteria Help Plants Tolerate Abiotic Stress," *Trends in Plant Science* 14, no. 1 (2009): 1–4.

25. Connell and Slatyer, "Mechanisms of Succession," 1123–1124.

26. Zwolinski, "Fire Effects on Vegetation and Succession," 21.

27. Vitousek et al., "Nitrogen Availability," 233; Eugene F. Kelly, Oliver A. Chadwick, and Thomas E. Hilinski, "The Effect of Plants on Mineral Weathering," *Biogeochemistry* 42 (1998): 21–53; Angela Hodge, "Root Decisions," *Plant, Cell & Environment* 32 (2009): 628–640.

28. Julie Sloan Denslow, "Patterns of Plant Species Diversity during Succession under Different Disturbance Regimes," *Oecologia* 46, no. 1 (1980): 18–21.

29. Knight and Chase, "Ecological Succession," R926; Vitousek et al., "Nitrogen Availability," 233.

30. Vitousek et al., "Nitrogen Availability," 230.

31. Connell and Slatyer, "Mechanisms of Succession"; Denslow, "Patterns of Plant Species Diversity."

32. Denslow, "Patterns of Plant Species Diversity," 18.

33. Vitousek et al., "Nitrogen Availability," 230; Zwolinski, "Fire Effects on Vegetation and Succession," 21–22.

34. The terms alpha and beta diversity, together with a third term, gamma diversity, were first introduced by R. H. Whittaker in 1960, in "Vegetation of the Siskiyou

Mountains, Oregon and California," *Ecological Monographs* 30 (1960): 279–338. See also Christopher M. Swan, Anna Johnson, and David J. Nowak, "Differential Organization of Taxonomic and Functional Diversity in an Urban Woody Plant Metacommunity," *Applied Vegetation Science* 20 (2017): 7–17.

35. Swan et al., "Differential Organization," 8.

36. Denslow, "Patterns of Plant Species Diversity," 18.

37. Swan et al., "Differential Organization," 10.

38. Sheikh Rabbi, Matthew K. Tighe, Richard J. Flavel, et al., "Plant Roots Redesign the Rhizosphere to Alter the Three-Dimensional Physical Architecture and Water Dynamics," *New Phytologist* 219, no. 2 (2018): 542–550.

39. Jan K. Schjoerring, Ismail Cakmak, and Philip J. White, "Plant Nutrition and Soil Fertility: Synergies for Acquiring Global Green Growth and Sustainable Development," *Plant and Soil* 434 (2019): 1–6; Adnan Noor Shah, Mohsin Tanveer, Babar Shahzad, et al., "Soil Compaction Effects on Soil Health and Crop Productivity: An Overview," *Environmental Science and Pollution Research* 24 (2017): 10056–10067.

40. Rabbi et al., "Plant Roots Redesign," 542; Debbie S. Feeney, John W. Crawford, Tim Daniell, et al., "Three-dimensional Microorganization of the Soil–Root–Microbe System," *Microbial Ecology* 52, no. 1 (2006): 151–158.

41. Kerry L. Metlen, Erik T. Aschehoug, and Ragan M. Callaway, "Plant Behavioural Ecology: Dy-

namic Plasticity in Secondary Metabolites," *Plant, Cell & Environment* 32, no. 6 (2009): 641–653.

42. Rabbi et al., "Plant Roots Redesign," 542; Feeney et al., "Three-dimensional Microorganization."

43. Dayakar V. Badri and Jorge M. Vivanco, "Regulation and Function of Root Exudates," *Plant, Cell & Environment,* 32, no. 6 (2009): 666–681; Metlen, Aschehoug, and Callaway, "Plant Behavioural Ecology."

44. Rabbi et al., "Plant Roots Redesign," 543.

45. D. B. Read, A. G. Bengough, P. J. Gregory, et al., "Plant Roots Release Phospholipid Surfactants That Modify the Physical and Chemical Properties of Soil," *New Phytologist* 157, no. 2 (2003): 315–326.

46. Read et al., "Plant Roots Release Phospholipid Surfactants," 316.

47. Ergosterol is a fungal-specific sterol found in the cell membranes of fungi that functions to maintain cell membrane permeability. It is a biomarker that is often quantified to estimate the biomass of mycorrhizal fungi association with plant roots or soil samples; Yongqiang Zhang, and Rajini Rao, "Beyond Ergosterol: Linking pH to Antifungal Mechanisms," *Virulence* 1, no. 6 (2010): 551–554.

48. The glycoprotein glomalin is an organic compound rich in carbon and nitrogen that is produced by arbuscular mycorrhizal fungi. It is released into the rhizosphere and alters soil properties such as aggregation and absorption of water; Karl Ritz and Iain M. Young, "Interactions between Soil Structure and Fungi," *Mycologist* 18, no. 2 (2004): 52–59; Matthias C. Rillig and

Peter D. Steinberg, "Glomalin Production by an Arbuscular Mycorrhizal Fungus: A Mechanism of Habitat Modification?" *Soil Biology and Biochemistry* 34, no. 9 (2002): 1371–1374.

49. Chang and Turner, "Ecological Succession in a Changing World," 506.

50. Lindsay Chaney and Regina S. Baucom, "The Soil Microbial Community Alters Patterns of Selection on Flowering Time and Fitness-related Traits in *Ipomoea purpurea*," *American Journal of Botany* 107, no. 2 (2020): 186–194; Chang and Turner, "Ecological Succession in a Changing World," 503.

51. James D. Bever, Thomas G. Platt, and Elise R. Morton, "Microbial Population and Community Dynamics on Plant Roots and Their Feedbacks on Plant Communities," *Annual Review of Microbiology* 66 (2012): 265–283; Tanya E. Cheeke, Chaoyuan Zheng, Liz Koziol, et al., "Sensitivity to AMF Species Is Greater in Late-Successional Than Early-Successional Native or Nonnative Grassland Plants," *Ecology* 100, no. 12 (2019): e02855; Liz Koziol and James D. Bever, "AMF, Phylogeny, and Succession: Specificity of Response to Mycorrhizal Fungi Increases for Late-Successional Plants," *Ecosphere* 7, no. 11 (2016): e01555; Liz Koziol and James D. Bever, "Mycorrhizal Feedbacks Generate Positive Frequency Dependence Accelerating Grassland Succession," *Journal of Ecology* 107, no. 2 (2019): 622–632.

52. Guillaume Tena, "Seeing the Unseen," *Nature Plants* 5 (2019): 647.

53. David P. Janos, "Mycorrhizae Influence Tropical Succession," *Biotropica* 12, no. 2 (1980): 56.

54. Janos, "Mycorrhizae Influence Tropical Succession," 58; Tereza Konvalinková and Jan Jansa, "Lights Off for Arbuscular Mycorrhiza: On Its Symbiotic Functioning under Light Deprivation," *Frontiers in Plant Science* 7 (2016): 782; Maki Nagata, Naoya Yamamoto, Tamaki Shigeyama, et al., "Red / Far Red Light Controls Arbuscular Mycorrhizal Colonization via Jasmonic Acid and Strigolactone Signaling," *Plant and Cell Physiology* 56, no. 11 (2015): 2100–2109; Maki Nagata, Naoya Yamamoto, Taro Miyamoto, et al., "Enhanced Hyphal Growth of Arbuscular Mycorrhizae by Root Exudates Derived from High R / FR Treated *Lotus japonicas*," *Plant Signaling & Behavior* 11, no. 6 (2016): e1187356.

55. Janos, "Mycorrhizae Influence Tropical Succession," 60.

56. Janos, "Mycorrhizae Influence Tropical Succession," 60.

57. Marzena Ciszak, Diego Comparini, Barbara Mazzolai, et al., "Swarming Behavior in Plant Roots," *PLOS One* 7, no. 1 (2012): e29759; Adrienne Maree Brown, *Emergent Strategy: Shaping Change, Changing Worlds* (Chico, CA: AK Press, 2017), 6.

58. Ciszak et al., "Swarming Behavior."

59. Dale Kaiser, "Bacterial Swarming: A Reexamination of Cell-Movement Patterns," *Current Biology* 17, no. 14 (2007): R561–R570.

60. Brown, *Emergent Strategy,* 12.

61. Ciszak et al., "Swarming Behavior."

62. Peter W. Barlow and Joachim Fisahn, "Swarms, Swarming and Entanglements of Fungal Hyphae and of Plant Roots," *Communicative & Integrative Biology* 6, no. 5 (2013): e25299-1.

63. Ciszak et al., "Swarming Behavior."

64. Barlow and Fisahn, "Swarms, Swarming, and Entanglements."

65. André Geremia Parise, Monica Gagliano, and Gustavo Maia Souza, "Extended Cognition in Plants: Is It Possible?" *Plant Signaling & Behavior* 15, no. 2 (2020): 1710661.

66. On prescribed fire, see Zwolinski, "Fire Effects on Vegetation and Succession," 18–24.

5. A Diverse Community

Epigraph: Andrea Wulf, *The Invention of Nature: Alexander von Humboldt's New World* (New York: Knopf, 2015), 125.

1. Cynthia C. Chang and Melinda D. Smith, "Resource Availability Modulates Above- and Below-Ground Competitive Interactions between Genotypes of a Dominant C4 Grass," *Functional Ecology* 28, no. 4 (2014): 1041–1051, 1042; David Tilman, *Resource Competition and Community Structure* (Princeton, NJ: Princeton University Press, 1982).

2. Philip O. Adetiloye, "Effect of Plant Populations on the Productivity of Plantain and Cassava Intercropping," *Moor Journal of Agricultural Research* 5, no. 1 (2004): 26–32; Long Li, David Tilman, Hans Lambers,

and Fu-Suo Zhang, "Plant Diversity and Overyielding: Insights from Belowground Facilitation of Intercropping in Agriculture," *New Phytologist* 203, no. 1 (2014): 63–69; Zhi-Gang Wang, Xin Jin, Xing-Guo Bao, et al., "Intercropping Enhances Productivity and Maintains the Most Soil Fertility Properties Relative to Sole Cropping," *PLOS One* 9 (2014): e113984.

3. Li et al., "Plant Diversity and Overyielding."

4. Venida S. Chenault, "Three Sisters: Lessons of Traditional Story Honored in Assessment and Accreditation," *Tribal College* 19, no. 4 (2008): 15–16; Robin Wall Kimmerer, *Braiding Sweetgrass: Indigenous Wisdom, Scientific Knowledge and the Teachings of Plants* (Minneapolis, MN: Milkweed Editions, 2015), 132.

5. Kimmerer, *Braiding Sweetgrass,* 128–140; K. Kris Hirst, "The Three Sisters: The Traditional Intercropping Agricultural Method," *ThoughtCo,* May 30, 2019, https://www.thoughtco.com/three-sisters-american-farming-173034.

6. Kimmerer, *Braiding Sweetgrass,* 131.

7. Kimmerer, *Braiding Sweetgrass,* 130.

8. Adetiloye, "Effect of Plant Populations on the Productivity of Plantain and Cassava Intercropping"; P. O. Aiyelari, A. N. Odede, and S. O. Agele, "Growth, Yield and Varietal Responses of Cassava to Time of Planting into Plantain Stands in a Plantain / Cassava Intercrop in Akure, South-West Nigeria," *Journal of Agronomy Research* 2, no. 2 (2019): 1–16.

9. Kimmerer, *Braiding Sweetgrass,* 131; Abdul Rashid War, Michael Gabriel Paulraj, Tariq Ahmad, et al.,

"Mechanisms of Plant Defense against Insect Herbivores," *Plant Signaling & Behavior* 7, no. 10 (2012): 1306–1320.

10. Kimmerer, *Braiding Sweetgrass,* 140.

11. Kimmerer, *Braiding Sweetgrass,* 132.

12. Lindsay Chaney and Regina S. Baucom, "The Soil Microbial Community Alters Patterns of Selection on Flowering Time and Fitness-related Traits in *Ipomoea purpurea,*" *American Journal of Botany* 107, no. 2 (2020): 186–194; Jennifer A. Lau and Jay T. Lennon, "Evolutionary Ecology of Plant–Microbe Interactions: Soil Microbial Structure Alters Selection on Plant Traits," *New Phytologist* 192, no. 1 (2011): 215–224; Marcel G. A. Van Der Heijden, Richard D. Bardgett, and Nico M. Van Straalen, "The Unseen Majority: Soil Microbes as Drivers of Plant Diversity and Productivity in Terrestrial Ecosystems," *Ecology Letters* 11, no. 3 (2008): 296–310.

13. Kimmerer, *Braiding Sweetgrass,* 133; Catherine Bellini, Daniel I. Pacurar, and Irene Perrone, "Adventitious Roots and Lateral Roots: Similarities and Differences," *Annual Review of Plant Biology* 65 (2014): 639–666.

14. Angela Hodge, "The Plastic Plant: Root Responses to Heterogeneous Supplies of Nutrients," *New Phytologist* 162, no. 1 (2004): 9–24.

15. Kimmerer, *Braiding Sweetgrass,* 140.

16. Henrik Hartmann and Susan Trumbore, "Understanding the Roles of Nonstructural Carbohydrates in Forest Trees—From What We Can Measure to What We Want to Know," *New Phytologist* 211, no. 2 (2016): 386–403.

17. Kimmerer, *Braiding Sweetgrass,* 133; Janet I. Sprent, "Global Distribution of Legumes," in *Legume Nodulation: A Global Perspective* (Oxford: Wiley-Blackwell, 2009), 35–50; Jungwook Yang, Joseph W. Kloepper, and Choong-Min Ryu, "Rhizosphere Bacteria Help Plants Tolerate Abiotic Stress," *Trends in Plant Science* 14, no. 1 (2009): 1–4.

18. Tamir Klein, Rolf T. W. Siegwolf, and Christian Körner, "Belowground Carbon Trade among Tall Trees in a Temperate Forest," *Science* 352, no. 6283 (2016): 342–344.

19. Cyril Zipfel and Silke Robatzek, "Pathogen-Associated Molecular Pattern-Triggered Immunity: *Veni, Vidi . . . ?*" *Plant Physiology* 154, no. 2 (2010): 551–554.

20. Kevin R. Bairos-Novak, Maud C. O. Ferrari, and Douglas P. Chivers, "A Novel Alarm Signal in Aquatic Prey: Familiar Minnows Coordinate Group Defences against Predators through Chemical Disturbance Cues," *Journal of Animal Ecology* 88, no. 9 (2019): 1281–1290.

21. Michiel van Breugel, Dylan Craven, Hao Ran Lai, et al., "Soil Nutrients and Dispersal Limitation Shape Compositional Variation in Secondary Tropical Forests across Multiple Scales," *Journal of Ecology* 107, no. 2 (2019): 566–581.

22. Robin Wall Kimmerer, "Weaving Traditional Ecological Knowledge into Biological Education: A Call to Action," *BioScience* 52, no. 5 (2002): 432–438.

23. Chenault, "Three Sisters."

24. See Kimmerer, *Braiding Sweetgrass,* 134.

25. Kimmerer, *Braiding Sweetgrass;* Jayalaxshmi Mistry and Andrea Berardi, "Bridging Indigenous and Scientific Knowledge," *Science* 352, no. 6291 (2016): 1274–1275.

26. Robin Wall Kimmerer, "The Intelligence in All Kinds of Life," *On Being with Krista Tippett,* original broadcast February 25, 2016, https://onbeing.org/programs/robin-wall-kimmerer-the-intelligence-in-all-kinds-of-life-jul2018/.

27. Joseph A.Whittaker and Beronda L. Montgomery, "Cultivating Institutional Transformation and Sustainable STEM Diversity in Higher Education through Integrative Faculty Development," *Innovative Higher Education* 39, no. 4 (2014): 263–275.

28. Whittaker and Montgomery, "Cultivating Institutional Transformation."

29. Kimmerer, *Braiding Sweetgrass,* 132.

30. Kimmerer, *Braiding Sweetgrass,* 58.

31. For examples of the role of cultural competence in promoting successful outcomes in collaboration, see Stephanie M. Reich and Jennifer A. Reich, "Cultural Competence in Interdisciplinary Collaborations: A Method for Respecting Diversity in Research Partnerships," *American Journal of Community Psychology* 38, no. 1–2 (2006): 51–62.

32. Joseph A. Whittaker and Beronda L. Montgomery, "Cultivating Diversity and Competency in STEM: Challenges and Remedies for Removing Virtual Barriers to Constructing Diverse Higher Education Communities of Success," *Journal of Undergraduate Neuroscience Education* 11, no. 1 (2012): A44–A51; Kim Parker, Rich Morin, and Juliana Menasce Horowitz, "Looking to the Future, Public Sees an America in Decline on Many Fronts," Pew Research Center,

March 2019), ch. 3, "Views of Demographic Changes," https://www.pewsocialtrends.org/wp-content/uploads/sites/3/2019/03/US-2050_full_report-FINAL.pdf.

6. A Plan for Success

Epigraph: Dawna Markova, *I Will Not Die an Unlived Life: Reclaiming Purpose and Passion* (Berkeley, CA: Conari Press, 2000), 1.

1. Cynthia C. Chang and Melinda D. Smith, "Resource Availability Modulates Above- and Below-ground Competitive Interactions between Genotypes of a Dominant C_4 Grass," *Functional Ecology* 28, no. 4 (2014): 1041–1051.

2. Jannice Friedman and Matthew J. Rubin, "All in Good Time: Understanding Annual and Perennial Strategies in Plants," *American Journal of Botany* 102, no. 4 (2015): 497–499.

3. Diederik H. Keuskamp, Rashmi Sasidharan, and Ronald Pierik, "Physiological Regulation and Functional Significance of Shade Avoidance Responses to Neighbors," *Plant Signaling & Behavior* 5, no. 6 (2010): 655–662.

4. Katherine M. Warpeha and Beronda L. Montgomery, "Light and Hormone Interactions in the Seed-to-Seedling Transition," *Environmental and Experimental Botany* 121 (2016): 56–65.

5. Lourens Poorter, "Are Species Adapted to Their Regeneration Niche, Adult Niche, or Both?" *American Naturalist* 169, no. 4 (2007): 433–442.

6. Anders Forsman, "Rethinking Phenotypic Plasticity and Its Consequences for Individuals, Populations and Species," *Heredity* 115 (2015): 276–284; Robert Muscarella, María Uriarte, Jimena Forero-Montaña, et al., "Life-history Trade-offs during the Seed-to-Seedling Transition in a Subtropical Wet Forest Community," *Journal of Ecology* 101, no. 1 (2013): 171–182; Warpeha and Montgomery, "Light and Hormone Interactions."

7. Carl Procko, Charisse Michelle Crenshaw, Karin Ljung, et al., "Cotyledon-generated Auxin Is Required for Shade-induced Hypocotyl Growth in *Brassica rapa,*" *Plant Physiology* 165, no. 3 (2014): 1285–1301; Chuanwei Yang and Lin Li, "Hormonal Regulation in Shade Avoidance," *Frontiers in Plant Science* 8 (2017): 1527.

8. Taylor S. Feild, David W. Lee, and N. Michele Holbrook, "Why Leaves Turn Red in Autumn. The Role of Anthocyanins in Senescing Leaves of Red-Osier Dogwood," *Plant Physiology* 127, no. 2 (2001): 566–574; Bertold Mariën, Manuela Balzarolo, Inge Dox, et al., "Detecting the Onset of Autumn Leaf Senescence in Deciduous Forest Trees of the Temperate Zone," *New Phytologist* 224, no. 1 (2019): 166–176; Edward J. Primka and William K. Smith, "Synchrony in Fall Leaf Drop: Chlorophyll Degradation, Color Change, and Abscission Layer Formation in Three Temperate Deciduous Tree Species," *American Journal of Botany* 106, no. 3 (2019): 377–388.

9. Some of the brightly colored pigments already were accumulated prior to autumn. Yet, it appears that energy

is invested in synthesizing additional anthocyanins at a time when it would seem prudent to limit energy spent on making new compounds because of their role in screening plant cells from phototoxicity during degreening; Feild et al., "Why Leaves Turn Red in Autumn"; Primka and Smith, "Synchrony in Fall Leaf Drop."

10. Monika A. Gorzelak, Amanda K. Asay, Brian J. Pickles, and Suzanne W. Simard, "Interplant Communication through Mycorrhizal Networks Mediates Complex Adaptive Behaviour in Plant Communities," *AoB Plants* 7 (2015): plv050.

11. Gorzelak et al., "Interplant Communication through Mycorrhizal"; David Robinson and Alastair Fitter, "The Magnitude and Control of Carbon Transfer between Plants Linked by a Common Mycorrhizal Network," *Journal of Experimental Botany* 50, no. 330 (1999): 9–13.

12. David P. Janos, "Mycorrhizae Influence Tropical Succession," *Biotropica* 12, no. 2 (1980): 56–64; Leanne Philip, Suzanne Simard, and Melanie Jones, "Pathways for Below-ground Carbon Transfer between Paper Birch and Douglas-fir Seedlings," *Plant Ecology & Diversity* 3, no. 3 (2010): 221–233.

13. Tamir Klein, Rolf T. W. Siegwolf, and Christian Körner, "Belowground Carbon Trade among Tall Trees in a Temperate Forest," *Science* 352, no. 6283 (2016): 342–344.

14. Peng-Jun Zhang, Jia-Ning Wei, Chan Zhao, et al., "Airborne Host–Plant Manipulation by Whiteflies via an Inducible Blend of Plant Volatiles," *Proceedings of the National Academy of Sciences* 116, no. 15 (2019): 7387–7396.

15. Sarah Courbier and Ronald Pierik, "Canopy Light Quality Modulates Stress Responses in Plants," *iScience* 22 (2019): 441–452.

16. Scott Hayes, Chrysoula K. Pantazopoulou, Kasper van Gelderen, et al., "Soil Salinity Limits Plant Shade Avoidance," *Current Biology* 29, no. 10 (2019): 1669–1676; Wouter Kegge, Berhane T. Weldegergis, Roxina Soler, et al., "Canopy Light Cues Affect Emission of Constitutive and Methyl Jasmonate-induced Volatile Organic Compounds in *Arabidopsis thaliana,*" *New Phytologist* 200, no. 3 (2013): 861–874.

17. Beronda L. Montgomery, "Planting Equity: Using What We Know to Cultivate Growth as a Plant Biology Community," *Plant Cell* 32, no. 11 (2020): 3372–3375.

18. I use the term "minoritized" for people or groups who "as a result of social constructs have less power or representation compared to other members or groups in society"; the term "minority" can simply indicate being smaller in number, rather than reflecting a systematic structure related to histories of oppression, exclusion, or other inequities. See I. E. Smith, "Minority vs. Minoritized: Why the Noun Just Doesn't Cut It," *Odyssey,* September 2, 2016, https://www.theodysseyonline.com/minority-vs-minoritize.

19. Emma D. Cohen and Will R. McConnell, "Fear of Fraudulence: Graduate School Program Environments and the Impostor Phenomenon," *Sociological Quarterly* 60, no. 3 (2019): 457–478; Mind Tools Content Team, "Impostor Syndrome: Facing Fears of Inad-

equacy and Self-Doubt," *Mindtools,* https://www.mindtools.com/pages/article/overcoming-impostor-syndrome.htm; Sindhumathi Revuluri, "How to Overcome Impostor Syndrome," *Chronicle of Higher Education,* October 4, 2018, https://www.chronicle.com/article/How-to-Overcome-Impostor/244700.

20. Beronda L. Montgomery, "Mentoring as Environmental Stewardship," *CSWEP News* 2019, no. 1 (2019): 10–12.

21. Montgomery, "Mentoring as Environmental Stewardship."

22. Angela M. Byars-Winston, Janet Branchaw, Christine Pfund, et al., "Culturally Diverse Undergraduate Researchers' Academic Outcomes and Perceptions of Their Research Mentoring Relationships," *International Journal of Science Education* 37, no. 15 (2015): 2533–2553; Christine Pfund, Christine Maidl Pribbenow, Janet Branchaw, et al., "The Merits of Training Mentors," *Science* 311, no. 5760 (2006): 473–474; Christine Pfund, Stephanie C. House, Pamela Asquith, et al., "Training Mentors of Clinical and Translational Research Scholars: A Randomized Controlled Trial," *Academic Medicine* 89, no. 5 (2014): 774–782; Christine Pfund, Kimberly C. Spencer, Pamela Asquith, et al., "Building National Capacity for Research Mentor Training: An Evidence-Based Approach to Training the Trainers," *CBE-Life Sciences Education* 14, no. 2 (2015): ar24.

23. Center for the Improvement of Mentored Experiences in Research, https://cimerproject.org/#/; National

Research Mentoring Network, https://nrmnet.net/; Becky Wai-Ling Packard, mentoring resources, n.d., https://commons.mtholyoke.edu/beckypackard/resources/.

24. Recent research and discussion have highlighted the need for culturally relevant practices in mentoring and leadership. Such practices recognize that individuals come from different backgrounds, with distinct cultural norms and practices. Mentors and leaders often have to increase their cultural competence to effectively support individuals from a broad range of different cultures; Torie Weiston-Serdan, *Critical Mentoring: A Practical Guide* (Sterling, VA: Stylus, 2017), 44; Angela Byars-Winston, "Toward a Framework for Multicultural STEM-Focused Career Interventions," Career *Development Quarterly* 62, no. 4 (2014): 340–357; Beronda L. Montgomery and Stephani C. Page, "Mentoring beyond Hierarchies: Multi-Mentor Systems and Models," Commissioned Paper for National Academies of Sciences, Engineering, and Medicine Committee on Effective Mentoring in STEMM (2018), https://www.nap.edu/resource/25568/Montgomery%20and%20Page%20-%20Mentoring.pdf.

25. Weiston-Serdan, *Critical Mentoring,* 44; see also Joseph A. Whittaker and Beronda L. Montgomery, "Cultivating Diversity and Competency in STEM: Challenges and Remedies for Removing Virtual Barriers to Constructing Diverse Higher Education Communities of Success," *Journal of Undergraduate Neuroscience Education* 11, no. 1 (2012): A44–A51.

26. Betty Neal Crutcher, "Cross-Cultural Mentoring: A Pathway to Making Excellence Inclusive," *Liberal Education* 100, no. 2 (2014): 26.

27. Weiston-Serdan, *Critical Mentoring,* 14.

28. George C. Banks, Ernest H. O'Boyle Jr., Jeffrey M. Pollack, et al., "Questions about Questionable Research Practices in the Field of Management: A Guest Commentary," *Journal of Management* 42, no. 1 (2016): 5–20; Ferrie C. Fang and Arturo Casadevall, "Competitive Science: Is Competition Ruining Science?" *Infection and Immunity* 83, no. 4 (2015): 1229–1233; Shina Caroline Lynn Kamerlin, "Hypercompetition in Biomedical Research Evaluation and Its Impact on Young Scientist Careers," *International Microbiology* 18, no. 4 (2015): 253–261; Beronda L. Montgomery, Jualynne E. Dodson, and Sonya M. Johnson, "Guiding the Way: Mentoring Graduate Students and Junior Faculty for Sustainable Academic Careers," *SAGE Open* 4, no. 4 (2014): doi: 10.1177/2158244014558043.

Conclusion

Epigraph: Monica Gagliano, *Thus Spoke the Plant: A Remarkable Journey of Groundbreaking Scientific Discoveries and Personal Encounters with Plants* (Berkeley, CA: North Atlantic Books, 2018), 93.

1. Sonia E. Sultan, "Developmental Plasticity: Reconceiving the Genotype," *Interface Focus* 7, no. 5 (2017): 20170009, 3.

2. Monica Gagliano, Michael Renton, Martial Depczynski, and Stefano Mancuso, "Experience Teaches Plants to Learn Faster and Forget Slower in Environments Where It Matters," *Oecologia* 175, no. 1 (2014): 63–72; Evelyn L. Jensen, Lawrence M. Dill, and James F. Cahill Jr., "Applying Behavioral-Ecological Theory to Plant Defense: Light-dependent Movement in *Mimosa pudica* Suggests a Trade-off between Predation Risk and Energetic Reward," *American Naturalist* 177, no. 3 (2011): 377–381; Franz W. Simon, Christina N. Hodson, and Bernard D. Roitberg, "State Dependence, Personality, and Plants: Light-foraging Decisions in *Mimosa pudica* (L.)," *Ecology and Evolution* 6, no. 17 (2016): 6301–6309.

3. Beronda L. Montgomery, "How I Work and Thrive in Academia—From Affirmation, Not for Affirmation," Being Lazy and Slowing Down Blog, September 30, 2019, https://lazyslowdown.com/how-i-work-and-thrive-in-academia-from-affirmation-not-for-affirmation/.

4. Beronda L. Montgomery, "Academic Leadership: Gatekeeping or Groundskeeping?" *Journal of Values-Based Leadership* 13, no. 2 (2020); Beronda L. Montgomery, "Mentoring as Environmental Stewardship," *CSWEP News* 2019, no. 1 (2019): 10–12.

5. Montgomery, "Academic Leadership"; Beronda L. Montgomery, "Effective Mentors Show up Healed," Beronda L. Montgomery website, December 5, 2019, http://www.berondamontgomery.com/mentoring/effective-mentors-show-up-healed/.

6. Andrew J. Dubrin, *Leadership: Researching Findings, Practice, and Skills,* 4th ed. (Boston: Houghton Mifflin, 2004).

7. Beronda L. Montgomery "Pathways to Transformation: Institutional Innovation for Promoting Progressive Mentoring and Advancement in Higher Education," Susan Bulkeley Butler Center for Leadership Excellence, Purdue University, Working Paper Series 1, no. 1, Navigating Careers in the Academy, 2018, 10–18, https://www.purdue.edu/butler/working-paper-series/docs/Inaugural%20Issue%20May2018.pdf.

8. Miller McPherson, Lynn Smith-Lovin, and, James M. Cook, "Birds of a Feather: Homophily in Social Networks," *Annual Review of Sociology* 27, no. 1 (2001): 415–444.

9. Montgomery, "Academic Leadership."

10. Szu-Fang Chuang, "Essential Skills for Leadership Effectiveness in Diverse Workplace Development," *Online Journal for Workforce Education and Development* 6, no. 1 (2013): 5; Katherine Holt and Kyoko Seki, "Global Leadership: A Developmental Shift for Everyone," *Industrial and Organizational Psychology* 5, no. 2 (2012): 196–215; Nhu TB Nguyen and Katsuhiro Umemoto, "Understanding Leadership for Cross-Cultural Knowledge Management," *Journal of Leadership Studies* 2, no. 4 (2009): 23–35; Joseph A. Whittaker and Beronda L. Montgomery, "Cultivating Institutional Transformation and Sustainable STEM Diversity in Higher Education through Integrative Faculty Development," *Innovative Higher Education* 39, no. 4 (2014): 263–275; Joseph A.

Whittaker, Beronda L. Montgomery, and Veronica G. Martinez Acosta, "Retention of Underrepresented Minority Faculty: Strategic Initiatives for Institutional Value Proposition Based on Perspectives from a Range of Academic Institutions," *Journal of Undergraduate Neuroscience Education* 13, no. 3 (2015): A136–A145; Torie Weiston-Serdan, *Critical Mentoring: A Practical Guide* (Sterling, VA: Stylus, 2017).

11. Stephanie M. Reich and Jennifer A. Reich, "Cultural Competence in Interdisciplinary Collaborations: A Method for Respecting Diversity in Research Partnerships," *American Journal of Community Psychology* 38, no. 1 (2006): 51–62.

12. Montgomery, "Academic Leadership."

13. Montgomery, "Mentoring as Environmental Stewardship."

14. Montgomery, "Academic Leadership."

Acknowledgments

To say that this book is a love story to plants does not accurately capture what it represents for me. I am truly appreciative that I've learned about reciprocity from plants. I'm thankful for members of my scientific community, who have shared knowledge about, enthusiasm for, and ongoing intrigue with plants with me over the decades.

I'd like to thank the supportive editorial team at Harvard University Press, including the tireless efforts of Janice Audet, who inspired me to dream that this project could be possible, and Louise Robbins, who served as an exemplary caretaker.

Portions of Chapter 5 were first published as "Three Sisters and Integrative Faculty Development," *Plant Science Bulletin* 63, no. 2 (2017): 78–85. Parts of Chapter 6 first appeared in "From Deficits to Possibilities," *Public Philosophy Journal* 1, no. 1 (2018). I am grateful to those publications for giving me the opportunity to present this early work.

My progress on this book was propelled by the phenomenal support that I have found in multiple writing spaces: the Faculty Writing Spaces and writing retreats by the Diversity Research Network, as well as the awe-inspiring space and support at Easton's Nook under the diligent care of sisters Jacquie and Nadine.

I'm grateful for my amazing family and legendary friends, who have always supported me. I'm not sure I have adequate words to thank my big sister, René. I've always said that your entry on the planet before me had to be by design. Although you were initially hired and quickly fired as my childhood scientific research assistant, you have persisted (and done so masterfully) as my best friend, as well as my first and most long-standing mentor. You have served in the role of mentor and guide valiantly, even when the task of doing so was big and complex. I've walked through nearly every challenge in life with you by my side (if not protectively in front of me), and I came through each one in large part due to your mentoring, wisdom, and endless patience. You've also always been co-celebrant for every triumph, including the writing of this book. Neither my life, nor this book, would be what it is without you!

Finally, of all the things that I aspired to do and do well, mothering Nicolas has always been my greatest priority, my ultimate joy! Every good thing that you are has been a gift to me. Thank you, Nicolas, for being a wonderful son, a brilliant and creative thinker, a generous and compassionate spirit, and unending inspiration to me in the bold, confident way you walk through life. Keep learning, keep giving, keep growing!

Index

Page numbers followed by the letter *f* denote figures.